THE EXTRACELLULAR MATRIX
FactsBook

Other books in the FactsBook Series:

A. Neil Barclay, Albertus D. Beyers, Marian L. Birkeland, Marion H. Brown,
Simon J. Davis, Chamorro Somoza, Alan F. Williams
The Leucocyte Antigen FactsBook

Robin E. Callard and Andy J. H. Gearing
The Cytokine FactsBook

Steve Watson and Steve Arkinstall
The G-Protein Linked Receptor FactsBook

Ed Conley
The Ion Channel FactsBook

Rod Pigott and Christine Power
The Adhesion Molecule FactsBook

THE EXTRACELLULAR MATRIX
FactsBook

Shirley Ayad
Ray Boot-Handford
Martin J. Humphries
Karl E. Kadler
Adrian Shuttleworth
School of Biological Sciences,
University of Manchester, Manchester, UK

Academic Press
Harcourt Brace & Company, Publishers
LONDON   SAN DIEGO   NEW YORK   BOSTON
SYDNEY   TOKYO   TORONTO

This book is printed on acid-free paper

ACADEMIC PRESS LIMITED
24–28 Oval Road
LONDON NW1 7DX

United States Edition published by
ACADEMIC PRESS INC.
San Diego, CA 92101

Copyright © 1994 by
ACADEMIC PRESS LIMITED

A catalogue record for this book is available from the British Library

ISBN 0–12–068910–3

Designed by Eric Drewery and Adrian Singer
Typeset by Columns Design and Production Services Ltd, Reading
Printed and bound in Great Britain by
Mackays of Chatham PLC, Chatham, Kent

Contents

Section I INTRODUCTORY MATERIAL

Section II THE EXTRACELLULAR MATRIX PROTEINS

This book is dedicated to David S. Jackson and Michael E. Grant for establishing and fostering matrix research in Manchester.

Preface

The authors would like to thank those colleagues who provided sequence and other information that was invaluable in the preparation of this book. These included: Scott Argraves, Mon-Li Chu, David Eyre, Tim Hardingham, Richard Mayne, Taini Pihlajaniemi, Francesco Ramirez, Jaro Sodek, Michel van der Rest, Marvin Tanzer and Arthur Veis. The authors would also like to acknowledge the funding of their own work by The Wellcome Trust, The Medical Research Council, and The Arthritis and Rheumatism Council. The authors would be grateful if they could be notified of any inaccuracies or omissions; the nature of the book means that these are inevitable but they can be corrected in future editions. Please send such correspondence to the Editor, Extracellular Matrix FactsBook, Academic Press, 24–28 Oval Road, London, NW1 7DX, UK.

From Top left: Martin Humphries, Adrian Shuttleworth, Ray Boot-Handford; bottom left: Karl Kadler, Shirley Ayad.

Abbreviations

CCP	Complement control protein
COL	Collagenous
COMP	Cartilage oligomeric matrix protein
COOH	Carboxyl
CS	Chondroitin sulphate
DEAE	Diethylaminoethyl
DS	Dermatan sulphate
EGF	Epidermal growth factor
EM	Electron microscopy
FACIT	Fibril-associated collagen with interrupted triple helix
FN	Fibronectin
FPLC	Fast protein liquid chromatography
GAG	Glycosaminoglycan
HA	Hyaluronan/hyaluronic acid
HS	Heparan sulphate
KS	Keratan sulphate
LDL	Low density lipoprotein
NC	Non-collagenous
N-linked	Asparagine-linked
NMR	Nuclear magnetic resonance
O-linked	Hydroxyl-linked
PARP	Proline- and arginine-rich protein
PG	Proteoglycan
SDS–PAGE	Polyacrylamide gel electrophoresis in sodium dodecyl sulphate
SLS	Segment-long-spacing
TGF	Transforming growth factor

INTRODUCTORY MATERIAL

Introduction

The purpose of this book is to collate and present key facts relating to the diverse group of macromolecules that assemble to form the extracellular matrix. An in-depth introduction to the structure and function of extracellular matrix is not within the remit of this book but several excellent reviews and monographs have been published recently on this topic and these are listed below. However, it is necessary to define the authors' perception of the extracellular matrix and our criteria for the inclusion and exclusion of various classes of macromolecules within this book.

The extracellular matrix is the substance which underlies all epithelia and endothelia, and surrounds all connective tissue cells providing mechanical support and physical strength to tissues, organs and the organism as a whole. In simple multicellular animals such as the hydra, the physical support for the whole organism may consist solely of a basement membrane. In more complex animals such as man, different types of extracellular matrix have evolved in connective tissues such as skin, bone, tendon, ligament and cartilage in addition to the ubiquitous basement membrane. The extracellular matrix should not be viewed as merely providing strength and physical support for tissues and organisms. It is now quite clear that this matrix exerts profound influences on both the behaviour (e.g. adherence, spreading and migration) and the pattern of gene expression of the cells in contact with it and, where identified, peptide sequences responsible for these effects are listed.

In deciding which groups of molecules to include or exclude from this book, we have taken the following view: to be included, a molecule must perform a structural role within the matrix and must be secreted, in its entirety, into the extracellular compartment. Accordingly, the following groups of molecules have been excluded: (1) membrane intercalated proteoglycans, since, although the extracellular domain of the molecule may well be performing a structural role within the extracellular matrix, a key function of this class of molecules may be to act as signal transducers across the plasma membrane and thus they would be more appropriately included in a volume dedicated to cell-surface receptors; and (2) extracellular molecules that become either intimately associated with the matrix or simply trapped but whose function is clearly not structural. Such molecules include metalloproteinases, growth factors and plasma proteins like albumin.

GENERAL FORMAT

Each entry in this book has been prepared using a common format. Factual information is allocated under a number of sub-headings; an introductory paragraph, and sections entitled Molecular structure, Isolation, Primary structure, Structural and functional sites, Gene structure and References. The criterion used to determine where particular information should be included in the entry reflects the level at which the information is defined. Where a particular structural or functional property has been shown to reside in a specific amino acid or series of amino acids, it is included under Structural and functional sites; where the property has been narrowed down to a specific region of the molecule but cannot be defined as a particular amino acid sequence, it is described under Molecular structure; and where the property cannot be assigned to a specific region of the molecule, it is described in the introductory paragraph. Accession number(s) are given under Primary structure to allow the reader to access sequences from databases. The particular numbers quoted usually reflect the first deposited sequence together with those for subsequent variant forms of the molecule. As a result, the codes vary in their origin and may refer to EMBL/Genbank, Swissprot or PIR databases. A complete listing of codes is rarely necessary and if needed these can usually be found by searching with the code that is given. References are given at the end of each entry and are given over primarily to review articles and major cloning and sequencing reports since there is insufficient space for supporting each factual statement with a reference.

In general, estimates of the relative mobilities of proteins in SDS–PAGE are not presented (except for collagen α chains in Table 1) because of the anomalous migration of most extracellular matrix molecules. Primary sequence information and database accession numbers are given for human molecules where possible, but only if the complete sequence is available. In addition, there are a number of known matrix molecules that were not completely sequenced at the time the book went to press, and these have therefore been left for future editions. All potential N-linked oligosaccharide attachment sequences are listed under Structural and functional sites. O-Linked oligosaccharide-attachment sites are listed only if established chemically.

DRAWINGS OF STRUCTURES

It is now well established that many protein molecules are built up from a series of independently folded polypeptide building blocks or repeats. Several of these repeats are found in a large number of proteins, suggesting that they might have arisen originally as autonomous polypeptides and that they were subsequently duplicated at the genomic level and incorporated into larger assemblies. Extracellular matrix macromolecules are good examples of proteins constructed in this way, since they contain many examples of protein repeats and in a number of cases, these repeats are catenated into mixed arrays.

The presence of independent repeats has implications not only for protein evolution, but also for function. It is wrong to assume that a repeat always serves the same function, and instead they should be viewed as structural units specialized for different functions in different proteins. Long arrays of repeats might provide a spacer function, particularly relevant for extracellular matrix molecules like fibrillin or laminin that form interconnected networks, or alternatively they may function as templates onto which specific recognition signals, or motifs, are grafted. One specialized example of this kind of function would be the evolution of tripeptide cell adhesive sequences within a small minority of fibronectin type III repeats.

In most cases, the repeat structure of extracellular matrix proteins has been determined by matrix-based database searching. The level of sequence identity is often in the range of 20–30% and the level of sequence similarity is slightly higher at 40–50%. Frequently, however, there are particular residues in a repeat that are crucial for structural integrity, and these are therefore more highly conserved. The presence of these residues also provides strong supportive evidence when assigning a particular repeat type to a polypeptide sequence.

For the purpose of this book, which is primarily concerned with facts about the extracellular matrix, it is important to present a perception of the repeat structure of extracellular matrix molecules. Since it is possible to represent this structure simply and visually in the form of a drawing that complements detailed sequence data, almost every entry in this book contains a view of the molecule concerned. The drawings contain some symbols that represent well-characterized protein repeats and others that portray regions of polypeptide that are less well defined in terms of structure. The latter are included simply to give as much information as possible about the substructure of the molecules. In addition, thick horizontal bars on some drawings indicate regions of the molecule which may undergo alternative splicing of pre-mRNA.

Thirty-three different symbols are used in the drawings in this book (including that for alternatively spliced sequences; see Figure 1). The repeats that are represented are as follows:

- Acidic. Short stretches of polypeptide containing continuous repeats of aspartic acid and/or glutamic acid residues. Structure unknown.
- Anaphylotoxin. Contain 30–40 amino acids including six cysteine residues. Found in complement components C3a, C4a and C5a and albumin. Structure unknown.
- Collagen triple helix. Variable size. Composed of repeating glycine–X–Y triplets (see page 8 for structural description).
- Elastin cross-link. Contain 10–30 amino acids including two lysine residues three or four amino acids apart that take part in inter-chain cross-linking. Structure unknown.
- Elastin hydrophobic. Contain 10–80 amino acids including a high content of hydrophobic residues (particularly alanine and valine). Structure unknown.
- Epidermal growth factor (EGF). Contain 30–80 amino acids. Six- or eight-cysteine versions are possible. Found in numerous proteins. The structure of EGF itself and the EGF repeat in factor IX have been determined [1,2].
- Fibrinogen β,γ. Contain approximately 200 amino acids. Structure unknown.
- Fibronectin type I. Contain 40–50 amino acids. Contain two intra-repeat disulphide bonds. Found in tissue plasminogen activator and factor XII. The structure of fibronectin type I-7 has been determined [3].
- Fibronectin type II/Kringle. Contain approximately 60 amino acids. Contain two intra-repeat disulphide bonds. Found in tissue plasminogen activator and plasminogen. The structure of the fibronectin type II repeat in a bovine seminal fluid protein has been determined [4].
- Fibronectin type III. Contain 90–100 amino acids. No intra-repeat disulphides. Found in numerous proteins. Several structures have been determined, e.g. RGD-containing repeats in fibronectin and tenascin [5,6]. These have a similar folding pattern to immunoglobulin domains, but have limited sequence homology.
- Gla. Region of polypeptide containing a number of γ-carboxyglutamic acid residues.
- Hemopexin. Contain 140–190 amino acids including three cysteine residues. Structure unknown.
- Heptad coiled-coil. A chain association domain that mediates the formation of

double- or triple-helical coiled coils. Heptad repeats contain non-polar residues at positions 1 and 4 and polar residues at 5 and 7 [7].

- Immunoglobulin (Ig). Contain 90–100 amino acids. Some Ig folds have an intra-repeat disulphide bond. Found in numerous proteins. Many structures have been determined (see ref. 8). All consist of a sandwich of two opposing sheets made up of β strands. Ig folds can be divided into V- and C-domains and further subdivisions are possible. C-domains contain three β strands on one face and four on the other; V-domains have extra sequence which adds two strands to the three-strand face (making five and four in total). All strands are connected by variable loops.
- Kunitz proteinase inhibitor. Contains approximately 50 amino acids including six cysteines. Structure unknown.
- Laminin G. Contain 160–230 amino acids including four cysteine residues. Structure unknown.
- LDL receptor. Contain 40–50 amino acids with no intra-repeat disulphide bonds. Structure unknown.
- Lectin. Contain 120–130 amino acids with two intra-repeat disulphide bonds. Also called C-type lectin because some proteins containing this repeat bind carbohydrate in a calcium-dependent manner. The structure of a C-type lectin domain from rat mannose-binding protein has been solved by X-ray crystallography and shown to contain two regions, one with irregular structure, the other containing both α helix and β sheet [9].
- Leucine-rich domain. Short stretches of polypeptide containing closely-spaced repeats of leucine residues. Structure unknown.
- Link protein. Contain 70–80 amino acids including four cysteine residues. Found in CD44. Structure unknown.
- Lysine/proline-rich. Region of polypeptide containing a high content of lysine and proline residues. Structure unknown.
- Ovomucoid. Contains approximately 55 amino acids including six cysteines. Found in several serine proteinase inhibitors. Structure unknown.
- PARP (proline- and arginine-rich protein). Contains approximately 210 amino acids. Structure unknown.
- Properdin. Contain approximately 50 amino acids including six cysteines. Found in complement components C6 and C9. Structure unknown.
- TGFβ1 receptor repeat. Contain 70–80 amino acids including eight cysteine residues. Structure unknown.
- Thrombospondin type 3 (TSP III). Contain 20–40 amino acids including two cysteines. These repeats also contain one or two copies of a 12-residue EF-hand-type cation-binding loop (coordination from residues 1, 3, 5, 7, 9 and 12). Structure otherwise unknown.
- Thyroglobulin. Contain 30–40 amino acids including several cysteine residues. Structure unknown.
- Tyrosine sulphate-rich. Region of polypeptide containing several sulphated tyrosine residues.
- von Willebrand factor A (vWF A). Contain 190–230 amino acids with no intra-repeat disulphide bonds. Found in many proteins. Structure unknown.
- von Willebrand factor B. Contain 25–35 amino acids including several cysteine residues. Structure unknown.

- von Willebrand factor C/procollagen. Contains approximately 70 amino acids including several cysteine residues. Structure unknown.
- von Willebrand factor D. Contain 270–290 amino acids including many cysteine residues. Structure unknown.

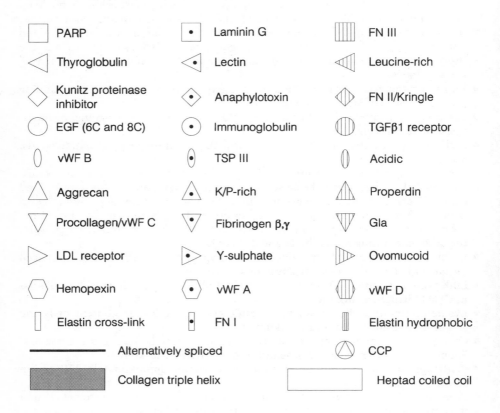

Figure 1. *Key to the artwork symbols used.*

References

[1] Cooke, R.M. et al (1987) The solution structure of human epidermal growth factor. Nature 327: 339–341.

[2] Handford, P.A. et al (1990) The first EGF-like domain from human factor IX contains high-affinity calcium binding site. EMBO J. 9: 475–480.

[3] Baron, M. et al (1990) Structure of the fibronectin type I module. Nature 345: 642–646.

[4] Constantine, K.L. et al (1992) Refined solution structure and ligand-binding properties of PDC-109 domain b. A collagen-binding type II domain. J. Mol. Biol. 223: 281–298.

[5] Leahy, D.J. et al (1992) Structure of a fibronectin type III domain from tenascin phased by MAD analysis of the selenomethionyl protein. Science 258: 987–991.

[6] Main, A.L. et al (1992) The three-dimensional structure of the tenth type III module of fibronectin: an insight into RGD-mediated interactions. Cell 71: 671–678.

[7] Cohen, C. and Parry, D. (1990) α-helical coiled coils and bundles: how to design an α-helical protein. Proteins: Struct. Func. Genet. 7: 1–15.

[8] Williams, A.F. and Barclay, A.N. (1988) The immunoglobulin superfamily – domains for cell surface recognition. Annu. Rev. Immunol. 6: 381–405.

[9] Weis, W.I. et al (1992) Structure of the calcium-dependent lectin domain from a rat mannose-binding protein determined by MAD phasing. Science 254: 1608–1615.

Further reading

Bork, P. (1992) The modular architecture of vertebrate collagens. FEBS Lett. 307: 49–54.

Neame, P.J., Young, C.N. and Treep, J.T. (1990) Isolation and primary structure of PARP, a 24-kDa proline- and arginine-rich protein from bovine cartilage closely related to the amino-terminal domain in collagen α1 (XI). J. Biol. Chem. 265: 20401–20408.

Extracellular matrices

Connective tissues join the other tissues of the body together. They take the stress of movement, maintain shape, and can be considered as a composite of insoluble fibres and soluble polymers. The principal fibres are collagen and elastin, while the soluble molecules include proteoglycans and glycoproteins. The structure (and hence function) of any connective tissue depends on the relative proportions of these constituent molecules. Those tissues that have to withstand large tensional force (such as tendon) tend to be particularly rich in fibrillar collagens, while a tissue that has to withstand compressive forces (such as cartilage) contains high levels of proteoglycans. The key features of the main classes of extracellular matrix are presented below.

COLLAGENS

The collagens constitute a highly specialized family of glycoproteins of which there are now at least 16 genetically distinct types encoded by at least 30 genes. Many have very complex structures and it is becoming increasingly difficult to define what is a collagen and what is not. At least three collagens comprise the protein cores of proteoglycans. Several proteins (e.g. the C1q component of complement and the enzyme acetylcholinesterase) contain the basic unit of a collagen – the triple helix – but for a protein to be classified as a collagen it must be an integral component of the extracellular matix.

The triple helix

The triple helix comprises three polypeptide (α) chains, each with a left-handed helical configuration, wound round each other to form a right-handed superhelix. Glycine occupies every third residue and this is an absolute requirement as glycine is the only residue with a small enough side-chain to fit into the centre of the triple helix without distorting it. Approximately 20–22% of the remaining X and Y residues in the repeating triplet $[Gly–X–Y]_n$ are, respectively, the imino acids proline and hydroxyproline. The hydroxyl group of hydroxyproline is essential for the formation of hydrogen bonds that stabilize the triple helix. Lysine and hydroxylysine residues in specific regions of both helical and non-helical regions are important in the formation of stable covalent cross-links within and between the collagen molecules in many of their supramolecular forms. Hydroxylysine residues are also potential sites for glycosylation with either galactose or glucosylgalactose. The hydroxylysine/lysine ratio as well as the degree of hydroxylysine glycosylation vary for the different collagen types and for the same collagen type in different tissues and with age. The triple-helical domains also vary in length for the different collagen types and can either be continuous or interrupted with non-helical domains.

The genetically distinct collagen types

The molecular configuration, supramolecular structure, approximate α-chain molecular weight (from the matrix form, estimated by SDS–PAGE) and tissue distribution for types I–XIV collagens are listed in Table 1. Similar data are not yet available for the more recently described types XV and XVI collagens.

Table 1. *The collagen family*

Type	Molecular configuration	Supramolecular structure	M_r x10^{-3} α-chain	Examples of tissue location
I	$[\alpha1(I)]_2\ \alpha2(I)$	Large-diameter 67 nm banded fibrils	95	Bone, cornea, skin, tendon,
	$[\alpha1(I)]_3$ trimer	67 nm banded fibrils	95	tumours, skin
II	$[\alpha1(II)]_3$	67 nm banded fibrils	95	Cartilage, vitreous
III	$[\alpha1(III)]_3$	Small-diameter 67 nm banded fibrils	95	Skin, aorta, uterus gut
IV	$[\alpha1(IV)]_2\ \alpha2(IV)$, plus $\alpha3(IV)$, $\alpha4(IV)$, $\alpha5(IV)$ chains	Non-fibrillar meshwork	170–180	Basement membranes
V*	$[\alpha1(V)]_2\ \alpha2(V)$ $[\alpha1(V)\alpha2(V)\alpha3(V)]$ $[\alpha1(V)]_3$	9 nm diameter non-banded fibrils	120–145	Placental tissue, bone, skin
VI	$[\alpha1(VI)\alpha2(VI)\alpha3(VI)]$ other forms?	5–10 nm diameter beaded microfibrils, 100 nm periodicity	140, 340	Uterus, skin, cornea, cartilage
VII	$[\alpha1(VII)]_3$	Anchoring fibrils	170	Amniotic membrane, skin, oesphagus
VIII	$[\alpha1(VIII)]_2\ \alpha2(VIII)$	Non-fibrillar hexagonal lattice	61	Descemet's membrane endothelial cells
IX	$[\alpha1(IX)\alpha2(IX)\alpha3(IX)]$	FACIT, non-fibrillar	68–115	Cartilage, vitreous
X	$[\alpha1(X)]_3$	Non-fibrillar, hexagonal lattice?	59	Calcifying cartilage
XI*	$[\alpha1(XI)\alpha2(XI)\alpha3(XI)]$	Fine fibrils similar to type V collagen	110–145	Cartilage, intervertebral disc
XII	$[\alpha1(XII)]_3$	FACIT, non-fibrillar	220, 340	Skin, tendon, cartilage
XIII	$\alpha1(XIII)$	?	62–67	Endothelial cells, epidermis
XIV	$[\alpha1(XIV)]_3$	FACIT, non-fibrillar	220, 340	Skin, tendon, cartilage

* Mixed Molecules comprising both type V and XI α-chains are known.

Collagen biosynthesis

The synthesis of collagen α chains follows the established pathway of other secretory proteins but involves several co- and post-translational modifications. These include the hydroxylation of proline and lysine, glycosylation of hydroxylysine, sulphation of tyrosine, disulphide bond formation and chain association, and the addition of complex carbohydrates and glycosaminoglycans (generally in non-collagenous domains) prior to secretion and deposition in the extracellular matrix. Additional modifications can occur in the extracellular matrix and depend on the collagen type and specific supramolecular structure formed.

The "classical" fibrillar collagens

Collagen types I, II and III constitute the classical fibril-forming collagens and account for 80–90% of all the collagen in the body. They are synthesized intracellularly as large precursor procollagens comprising a continuous triple helix, at each end of which are non-helical domains (propeptides). The propeptides are cleaved extracellularly by specific N- and C-proteinases during fibrillogenesis giving rise to the collagen monomers consisting of the triple helix (length 285 nm, diameter 1.4 nm) flanked by short non-helical regions. The monomers spontaneously assemble to form the fibril, each monomer staggered by 234 amino acid residues (a D unit 67 nm long) which results in maximal electrostatic and hydrophobic interactions between adjacent monomers and allows specific lysine/hydroxylysine residues in the helix and non-helical regions within the 0.4D overlap region to form stable covalent cross-links. The cross-links are formed by the oxidation of the non-helical ε-NH$_2$ group of lysine/hydroxylysine by the enzyme lysyl oxidase and the subsequent condensation of the resulting aldehyde with the ε-NH$_2$ group of the helical lysine/hydroxylysine residue to form a Schiff's base. These bifunctional cross-links condense with additional lysine/hydroxylysine or histidine residues on adjacent molecules to form the more mature stable trifunctional cross-links.

Type V and XI collagens are also classified as fibrillar collagens on the basis of their homology with type I–III collagens. All other collagens deviate considerably from this class of collagen in both structure and assembly within the extracellular matrix.

Recent studies indicate that there may be no such thing as a "type I, II or III collagen fibril". Research suggests that these major collagens can exist as heterotypic fibrils with I and III forming copolymers, and with type V and XI collagens copolymerized largely on the inside of type I and II collagen fibrils, respectively. Other collagens known as FACIT collagens (Fibril Associated Collagens with Interrupted Triple Helix), such as types IX, XII and XIV, associate with the surface of fibrils and modify their interactive properties.

Other collagen types

The interaction of fibrillar collagens with other collagens present either on the surface or in the interior of the fibril may alter their interactive capacity. This situation is likely to vary from tissue to tissue , and in the same tissue during development. In addition, there are a variety of collagenous molecules that can form three-dimensional networks (type IV), beaded filaments (type VI), anti-parallel dimers which form anchoring fibrils (type VII) and a hexagonal lattice (type VIII).

Nomenclature

All triple-helical collagenous (COL) domains and the non-helical, non-collagenous (NC) domains are numbered from the carboxy-terminal domain. In the case of the fibrillar collagens which are synthesized as larger precursor procollagen (pro-) forms, only the COL domains are numbered, the NC domains being the propeptides that are completely or partially cleaved on processing. The term pre-procollagen, previously denoting the procollagen form with attached signal sequence, will not be used for the complete primary structure: only the individual chain will be specified, e.g. $\alpha 1(I)$ chain.

PROTEOGLYCANS

Proteoglycans are a diverse family of molecules characterized by a core protein to which is attached one or more glycosaminoglycan (GAG) side-chains. These molecules appear to be distributed ubiquitously among animal cells, and participate in a variety of biological processes. Proteoglycans are found outside cells, intercalated in cell membranes and intracellularly in storage granules. This book is only concerned with those proteoglycans that are found in the extracellular matrix.

The heterogeneity of proteoglycan structure is a reflection not only of a variation in protein core, but also variation in the type and size of the GAG chains. Indeed, cells at different stages of development express different variants of the same proteoglycan. In recent times, proteoglycan nomenclature reflects the protein core and/or function of the molecule, whereas previously classification has been according to the chemical composition of the GAG chains. The glycosaminoglycan group of complex carbohydrates includes chondroitin sulphate (CS), dermatan sulphate (DS), heparan sulphate (HS), keratan sulphate (KS) and hyaluronan (HA). The characteristic feature of these molecules is that they are composed of a disaccharide repeat sequence of two different sugars; one of these is normally a hexuronate while the other is a hexosamine. Configurational variation in the disaccharide bonds and the position of sulphation leads to increased diversity in the chemical and physical properties of the chains.

Chondroitin sulphate

The basic repeating unit is formed from a glucuronic acid joined to N-acetylgalactosamine by $\beta 1–3$ linkage. Sulphate residues can be added to the 4 and 6 position of the amine residues, and are referred to as chondroitin-4-sulphate and chondroitin-6-sulphate. More commonly, a given chain exhibits stretches of 4-sulphation followed by 6-sulphation, and there are frequent occasions where neither residue in the disaccharide is sulphated or where both positions in an individual residue are sulphated.

Dermatan sulphate

This GAG has a similar structure to chondroitin sulphate except that some glucuronic acid residues are epimerized and converted to iduronic acid. Once formed, the iduronic acid residues may be sulphated at the 2 position. The glycosidic bond is also changed to α1–3. A CS chain with one or more iduronic acid residues is called dermatan sulphate.

Keratan sulphate

In the case of KS, the uronic acid residue is replaced by galactose. The repeat sequence is galactose joined to N-acetylglucosamine by a β1–4 linkage. Sulphation can occur on position 6 of each residue.

Heparan sulphate

The repeating sequence is glucuronic acid joined to N-acetylglucosamine by a β1–4 linkage. Both residues can be extensively sulphated at O- and N-positions, and the uronic acid can undergo epimerization.

Hyaluronan (hyaluronic acid)

Hyaluronan consists of an alternating polymer of glucuronic acid and N-acetylglucosamine joined by β1–3 linkage.

The high content of sulphate and the presence of uronic acid confer a large negative charge on glycosaminoglycans and permit them to associate with a large number of ligands by electrostatic interactions. Only a limited amount of information is available on sequences that confer specific binding.

Apart from hyaluronan, all the other GAG chains are found in tissues covalently attached to protein and function as proteoglycans. In general, the GAG chains are attached via a xylose residue linkage to specific serine residues in the protein core, the exception being keratan sulphate chains which are attached to protein via N- and O-linked glycosidic linkages to asparagine or serine/threonine, respectively.

Hyaluronan is a unique member of the GAG family since it functions *in vivo* as a free carbohydrate. It is polyanionic and large; a single molecule can have a molecular weight up to 10 million. HA assumes a randomly kinked, coil structure which occupies a large solution volume and endows solutions with high viscosity. The individual molecules can self-associate and form networks. A number of proteoglycans can associate with HA and form supramolecular aggregates in tissues. In developing tissues, HA can be the major structural macromolecule and supports both cell proliferation and migration. Our exclusion of HA from this book is in no way a reflection of the contribution of the molecule to matrix formation and function. Despite considerable developments in the structure, cellular interactions and nature of receptors, there is still very little functional data available relating to specific sites on the molecule.

Proteoglycans are diverse in both structural and functional terms. The highly acidic and hydrophilic GAG chains will have a major influence on tissue hydration and elasticity, and can show high affinity binding to a variety of ligands. Protein cores have also been shown to have specific interactions in the matrix. Developments in both protein and carbohydrate chemistry will help define the molecular basis for the diverse biological functions that these molecules participate in.

GLYCOPROTEINS

The simple picture of connective tissues as frameworks of insoluble fibrils and soluble polymers serves to highlight the importance of collagen fibres in resisting tensile stress and proteoglycans in associating with large amounts of water, swelling and resisting compressive forces. While these functions of the major structural elements of matrices are critical to many aspects of the physical properties of connective tissues, they take no account of the role of the large number of glycoproteins that are present in the ECM. A variety of extracellular glycoproteins have been isolated and the key feature of many of these molecules is their ability to interact not only with cells, but also with other macromolecules in the matrix.

Many of the matrix glycoproteins contain distinct and functionally active peptide domains that prescribe interactions with cell-surface receptors as well as other matrix molecules. In the case of molecules found in mineralized connective tissue, this may also involve interaction with the inorganic phase of bone or dentine. Some of the molecules form oligomers by disulphide bond formation or by non-covalent association. In addition, alternative splicing is a frequent finding producing families of closely related proteins. This heterogeneous group of molecules collectively contains carbohydrate covalently attached to the protein core which in some cases is O-linked, in others N-linked, and some contain both O- and N-linked carbohydrate. Phosphorylation of serine and threonine, and sulphation of tyrosine residues, are other post-translational modifications that might be found. A number of these molecules have been described as adhesive glycoproteins, the best studied of which is fibronectin.

The term "adhesive glycoproteins" includes such molecules as fibronectin, vitronectin, laminin, thrombospondin and von Willebrand factor which have several features in common. Distinct structural and functional sites are found in these molecules and some repetitive structural motifs (fibronectin type III and EGF repeats) are common to a number of adhesive molecules. Many possess short aspartate-containing sequences such as RGD, which mediate cellular adhesion through the integrin family of membrane receptors. This ability to interact with cells is clearly a key function, and we have chosen to include fibrinogen in this book because it also functions as a major adhesive molecule in blood, and where there has been damage to and leakage from the endothelium, it appears in the extracellular matrix and can influence fibroblast function.

In addition to the adhesive molecules, a number of other matrix glycoproteins have been characterized, particularly from skeletal tissue (cartilage, bone and dentine). One characteristic feature of many of the bone-associated molecules is their anionic nature. This fixed negative charge is a consequence of a variety of features; many are rich in the acidic amino acids (aspartic and glutamic), some contain stretches of consecutive aspartic acid residues (osteopontin), others stretches of consecutive glutamic acid residues (bone sialoprotein). We have chosen to present these as acidic-rich motifs. Other molecules contain γ-carboxyglutamic acid residues (osteocalcin and matrix Gla protein), and these have been shown as Gla-containing protein motifs. No Gla-recognition sequences have been shown. Sometimes the negative charge is provided by sialic acid residues on the oligosaccharide side chains (osteopontin, bone sialoprotein). A number of post-translational modifications also contribute to the anionic nature of the molecules, notably phosphorylation of serine (phosphoryns) and threonine, and sulphation of tyrosine residues (bone sialoprotein). This produces a variety of molecules that have been implicated in the process of calcification and calcium ion binding, although no specific functions have been demonstrated. In general these molecules are of low molecular weight (4–40 kDa), and tend to exist and function in monomeric form.

Apart from the adhesive glycoproteins and those isolated from skeletal tissues, another class of glycoproteins has been described, the members of which are associated with elastin and elastin deposition, notably fibrillin and MAGP. The best characterized of these is fibrillin, which has been clearly shown to be related to the incidence of Marfan's syndrome. A number of other molecules which have been implicated with elastin fibre formation have not been included, because full identification and structures are not available.

Glycoproteins of the extracellular matrix clearly have a number of important functions and are implicated in a variety of developmental and pathological changes in the extracellular matrix. The ability to influence cell behaviour by allowing attachment and migration of cells is clearly a major function of the adhesive glycoproteins. The ability to influence ion concentrations is a probable mode of action of a number of the skeletal glycoproteins, although some appear able to influence bone cell metabolism directly. Protein–protein interactions are also important in the functioning of some molecules, as evidenced by the ability of fibrillin to influence the deposition of elastin. The diversity and versatility of these molecules serves to highlight the importance of this class of molecule in the ECM.

Bibliography

General

Hay, E.D., ed. (1991) Cell Biology of the Extracellular Matrix, 2nd edition, Plenum Press, New York.

Mecham, R., ed. Biology of the Extracellular Matrix: A Series, Academic Press, New York. This series includes the following volumes;
  (1) Regulation of Matrix Accumulation, Mecham, R.P., ed.
  (2) Biology of Proteoglycans, Wight, T.N. and Mecham, R.P., eds.
  (3) Structure and Function of Collagen Types, Mayne, R. and Burgeson, R.E., eds.
  (4) Fibronectin, Mosher, D.F., ed.
  (5) Organisation and Assembly of Plant and Animal Extracellular Matrix, Adair,W.S. and Mecham, R.P., eds.
  (6) Extracellular Matrix Genes, Sandell, L.J. and Boyd, C.D., eds.

Royce, P.M. and Steinmann, B., eds (1993) Connective Tissue and its Heritable Disorders, Molecular, Genetic, and Medical Aspects, Wiley-Liss Inc., New York.

Collagens

Burgeson, R.E. and Nimni, M.E. (1992) Collagen types. Molecular structure and tissue distribution. Clin. Orthop. 282: 250–272.

Eyre, D.R. (1987) Collagen cross-linking amino acids. Methods Enzymol. 144: 115–139.

Kagan, H.M. (1986) Characterization and regulation of lysyl oxidase. In: Regulation of Matrix Accumulation: Biology of the Extracellular Matrix, Mecham, R.P., ed., Vol. 1, Academic Press, Orlando, pp. 321–398.

Kielty, C.M. et al (1993) The collagen family: structure, assembly and organization in the extracellular matrix. In: Connective Tissue and its Heritable Disorders. Molecular Genetics and Medical Aspects, Royce, P.M. and Steinmann, B., eds, Wiley-Liss, Inc., New York, pp. 103–147.

Kivirikko, K. and Myllyla, R. (1985) Post-translational processing of procollagens. Ann. N.Y. Acad. Sci. 460: 187–201.

Kuhn, K. (1987) The classical collagens: types I, II and III. In: Structure and Function of Collagen Types, Mayne, R. and Burgeson, R.E., eds, Academic Press, London, pp. 1–42.

Ramachandran, G.N. and Ramakrishnan, C. (1976) Molecular structure. In: Biochemistry of Collagen, Ramachandran, G.N. and Reddi, A.H., eds, Plenum Press, New York, pp. 45–84.

van der Rest, M. and Garrone, R. (1991) Collagen family of proteins. FASEB J. 5: 2814–2823.

Proteoglycans

Bourdon, M.A. et al (1987) Identification and synthesis of a recognition signal for the attachment of glycosaminoglycans to proteins. Proc. Natl. Acad. Sci. USA 84: 3194–3198.

Esko, J.D. (1991) Genetic analysis of proteoglycan structure, function and metabolism. Current Opinion Cell Biol. 3: 805–816.

Gallagher, J.T. (1989) The extended family of proteoglycans: social residents of the pericellular zone. Current Opinion Cell Biol. 1: 1201–1218.

Heinegard, D. and Oldberg, A. (1993) Glycosylated matrix proteins. In: Connective Tissue and its Heritable Disorders, Royce, P.M. and Steinman, B., eds, Wiley-Liss, New York, pp. 189–209.

Jackson, R.L. et al (1991) Glycosaminoglycans: molecular properties, protein interactions, and role in physiological processes. Physiol. Rev. 71: 481–539.

Kjellen, L. and Lindahl, U. (1991) Proteoglycans: Structure and interactions. Annu. Rev. Biochem. 60: 443–475.

Scott, J.E. (1992) Supramolecular organisation of extracellular matrix glycosaminoglycans in vitro and in the tissues. FASEB J. 6: 2639–2645.

Wight, T.N. et al (1992) The role of proteoglycans in cell adhesion, migration and proliferation. Current Opinion Cell Biol. 4: 793–801.

Glycoproteins

Bork, P. (1992) Mobile modules and motifs. Current Opinion Struct. Biol. 2: 413–421.

Heinegard, D. and Oldberg, A. (1989) Structure and biology of cartilage and bone matrix noncollagenous macromolecules. FASEB J. 3: 2042–2051.

Edelman, G.M. et al, eds, (1990) Morphoregulatory Molecules, Wiley & Sons, New York.

Mosher, D.F. et al (1992) Assembly of the extracellular matrix. Current Opinion Cell Biol. 4: 810–818.

Paulsson, M. (1992) Basement membrane proteins: Structure, assembly and cellular interactions. Crit. Rev. Biochem. Mol. Biol. 27: 93–127.

Schwarzbauer, J.E. (1991) Fibronectin: from gene to protein. Current Opinion Cell Biol. 3: 786–791.

Timpl, R. (1989) Structure and biological activity of basement membrane proteins. Eur. J. Biochem. 180: 487–502.

von der Mark, K. and Goodman, S. (1993) Adhesive glycoproteins. In: Connective Tissue and its Heritable Disorders, Royce, P.M. and Steinman, B., eds, Wiley-Liss, New York, pp. 211–236.

Yurchenco, P.D. and Schittny, J.C. (1990) Molecular architecture of basement membranes. FASEB J. 4: 1577–1590.

THE EXTRACELLULAR MATRIX PROTEINS

Aggrecan large aggregating CS-PG

Aggrecan is a large chondroitin sulphate proteoglycan (CS-PG) that accounts for about 10% of the dry weight of cartilage. Aggrecan interacts with hyaluronan (HA) via an HA-binding domain. The interaction has a dissociation constant of about 10^{-8} M and is additionally strengthened by the binding of link protein. Aggrecan is usually found as part of a large aggregate containing approximately 100 proteoglycan molecules per HA molecule. Aggrecan carries a large number of fixed negatively charged side-groups that result in high osmotic pressure in the tissue. It is generally accepted that the primary role of aggrecan is to swell and hydrate the framework of collagen fibrils in cartilage.

Molecular structure

Aggrecan comprises about 87% CS, 6% KS and 7% protein and has a molecular mass of 2.6×10^6 Da with a core protein of molecular mass approximately 220 kDa. By rotary shadowing electron microscopy and comparison of cDNAs of aggrecans from several species, the core protein is seen to have seven domains. The first three domains at the NH2-terminus comprise two globular domains called G1 and G2 (separated by an interglobule domain, IGD) which have structural similarities to the link protein of cartilage. G1 is a copy of the complete link protein sequence that can be divided into A (immunoglobulin repeat), B and B' subdomains (link repeats) and contains the hyaluronan binding domain (subdomain A), G2 is a copy of the COOH-terminal two-thirds of link protein and contains the B and B' subdomains (both link repeats). The next three domains contain sites for attachment of glycosaminoglycan (GAG). Keratan sulphate is the principal GAG in the fourth domain and CS is the principal GAG in the fifth and sixth domains. The seventh domain at the COOH-terminus contains a globule called G3 that is composed of a lectin repeat. There are at least three forms of aggrecan transcripts generated by alternative exon usage.

Alternatively spliced sequences are a 38 amino acid sequence inserted before the seventh domain that is highly homologous to an EGF repeat, and a complement control protein repeat that can be spliced out after the seventh domain. Aggrecan is cleaved by recombinant stromelysin between residues N341 and F342 that separates G1 from the remainder of the molecule. Evidence exists for a similar cleavage *in vivo* producing a non-aggregating form of aggrecan [1-3].

Isolation

Aggrecan is readily isolated from cartilage by extraction with 4 M guanidine-HCl/0.05 M sodium acetate pH 5.8 at 4°C, and is purified by sequential density gradient centrifugation and ion exchange chromatography [4]. A simple one-step purification procedure using 1.32–10% polyacrylamide gels has been described by Vilim and Krajickova [5].

Accession number

J05062; A39086

Primary structure

Ala	A	166	Cys	C	34	Asp	D	108	Glu	E	268
Phe	F	68	Gly	G	292	His	H	34	Ile	I	82
Lys	K	28	Leu	L	179	Met	M	13	Asn	N	30
Pro	P	219	Gln	Q	61	Arg	R	79	Ser	S	297
Thr	T	197	Val	V	186	Trp	W	25	Tyr	Y	49

Mol. wt (calc.) = 251 740   Residues = 2415

```
1     MTTLLWVFVT   LRVITAAVTV   TSDHDNSLS    VSIPQPSPLR   VLLGTSLTIP
51    CYFIDPMHPV   TTAPSTAPLA   PRIKWSRVSK   EKEVVLLVAT   EGRVRVNSAY
101   QDKVSLPNYP   AIPSDATLEV   QSLRSNDSGV   YRCEVMHGIE   DSEATLEVVV
151   KGIVFHYRAI   STRYTLDFDR   AQRACLQNSA   IIATPEQLQA   AYEDGFHQCD
201   AGWLADQTVR   YPIHTPREGC   YGDKDEFPGV   RTYGIRDTNE   TYDVYCFAEE
251   MEGEVFYATS   PEKFTFQEAA   NECRRLGARL   ATTGHVYLAW   QAGMDMCSAG
301   WLADRSVRYP   ISKARPNCGG   NLLGVRTVYV   HANQTGYPDP   SSRYDAICYT
351   GEDFVDIPEN   FFGVGGEEDI   TVQTVTWPDM   ELPLPRNITE   GEARGSVILT
401   VKPIFEVSPS   PLEPEEPFTF   APEIGATAFA   EVENETGEAT   RPWGFPTPGL
451   GPATAFTSED   LVVQVTAVPG   QPHLPGGVVF   HYRPGPTRYS   LTFEEAQQAC
501   PGTGAVIASP   EQLQAAYEAG   YEQCDAGWLR   DQTVRYPIVS   PRTPCVGDKD
551   SSPGVRTYGV   RPSTETYDVY   CFVDRLEGEV   FFATRLEQFT   FQEALEFCES
601   HNATATTGQL   YAAWSRGLDK   CYAGWLADGS   LRYPIVTPRP   ACGGDKPGVR
651   TVYLYPNQTG   LPDPLSRHHA   FCFRGISAVP   SPGEEEGGTP   TSPSGVEEWI
701   VTQVVPGVAA   VPVEEETTAV   PSGETTAILE   FTTEPENQTE   WEPAYTPVGT
751   SPLPGILPTW   PPTGAETEES   TEGPSATEVP   SASEEPSPSE   VPFPSEEPSP
801   SEEPFPSVRP   FPSVELFPSE   EPFPSKEPSP   SEEPSASEEP   YTPSPPEPSW
851   TELPSSGEES   GAPDVSGDFT   GSGDVSGHLD   FSGQLSGDRA   SGLPSGDLDS
901   SGLTSTVGSG   LTVESGLPSG   DEERIEWPST   PTVGELPSGA   EILEGSASGV
951   GDLSGLPSGE   VLETSASGVG   DLSGLPSGEV   LETTAPGVED   ISGLPSGEVL
1001  ETTAPGVEDI   SGLPSGEVLE   TTAPGVEDIS   GLPSGEVLET   TAPGVEDISG
1051  LPSGEVLETT   APGVEDISGL   PSGEVLETAA   PGVEDISGLP   SGEVLETAAP
1101  GVEDISGLPS   GEVLEIAAPG   VEDISGLPSG   EVLETAAPGV   EDISGLPSGE
1151  VLETAAPGVE   DISGLPSGEV   LETAAPGVED   ISGLPSGEVL   ETAAPGVEDI
1201  SGLPSGEVLE   TAAPGVEDIS   GLPSGEVLET   AAPGVEDISG   LPSGEVLETA
1251  APGVEDISGL   PSGEVLETAA   PGVEDISGLP   SGEVLETTAP   GVEEISGLPS
1301  GEVLETTAPG   VDEISGLPSG   EVLETTAPGV   EEISGLPSGE   VLETSTSAVG
1351  DLSGLPSGGE   VLEISVSGVE   DISGLPSGEV   VETSASGIED   VSELPSGEGL
1401  ETSASGVEDL   SRLPSGEEVL   EISASGFGDL   SGVPSGGEGL   ETSASEVGTD
1451  LSGLPSGREG   LETSASGAED   LSGLPSGKED   LVGSASGDLD   LGKLPSGTLG
1501  SGQAPETSGL   PSGFSGEYSG   VDLGSGPPSG   LPDFSGLPSG   FPTVSLVDST
1551  LVEVVTASTA   SELEGRGTIG   ISGAGEISGL   PSSELDISGR   ASGLPSGTEL
1601  SGQASGSPDV   SGEIPGLFGV   SGQPSGFPDT   SGETSGVTEL   SGLSSGQPGV
1651  SGEASGVLYG   TSQPFGITDL   SGETSGVPDL   SGQPSGLPGF   SGATSGVPDL
1701  VSGTTSGSGE   SSGITFVDTS   LVEVAPTTFK   EEEGLGSVEL   SGLPSGEADL
1751  SGKSGMVDVS   GQFSGTVDSS   GFTSQTPEFS   GLPSGIAEVS   GESSRAEIGS
1801  SLPSGAYYGS   GTPSSFPTVS   LVDRTLVESV   TQAPTAQEAG   EGPSGILELS
1851  GAHSGAPDMS   GEHSGFLDLS   GLQSGLIEPS   GEPPGTPYFS   GDFASTTNVS
1901  GESSVAMGTS   GEASGLPEVT   LITSEFVEGV   TEPTISQELG   QRPPVTHTPQ
1951  LFESSGKVST   AGDISGATPV   LPGSGVEVSS   VPESSSETSA   YPEAGFGASA
2001  APEASREDSG   SPDLSETTSA   FHEANLERSS   GLGVSGSTLT   FQEGEASAAP
2051  EVSGESTTTS   DVGTEAPGLP   SATPTASGDR   TEISGDLSGH   TSQLGVVIST
```

2101	SIPESEWTQQ	TQRPAETHLE	IESSSLLYSG	EETHTVETAT	SPTDASIPAS
2151	PEWKRESEST	AAAPARSCAE	EPCGAGTCKE	TEGHVICLCP	PGYTGEHCNI
2201	DQEVCEEGWN	KYQGHCYRHF	PDRETWVDAE	RRCREQQSHL	SSIVTPEEQE
2251	FVNNNAQDYQ	WIGLNDRTIE	GDFRWSDGHP	MQFENWRPNQ	PDNFFAAGED
2301	CVVMIWHEKG	EWNDVPCNYH	LPFTCKKGTV	ACGEPPVVEH	ARTFGQKKDR
2351	YEINSLVRYQ	CTEGFVQRHM	PTIRCQPSGH	WEEPRITCTD	PTTYKRRLQK
2401	RSSRHPRRSR	PSTAH			

Structural and functional sites

Signal peptide: 1–19
G1: 48–350
   Immunoglobulin repeat: 48–141
   Link repeats: 152–247, 253–349
IGD: 351–477
G2: 478–673
   Link repeats: 478–572, 579–673
KS attachment domain: 677–861
CS attachment domain 1: 864–1510
CS attachment domain 2: 1511–2162
EGF (6C) repeat: 2163–2200
Lectin repeat (G3): 2201–2329
CCP repeat: 2330–2391
Potential N-linked glycosylation sites: 126, 239, 333, 387, 434, 602, 657, 737, 1898

Gene structure

The human aggrecan gene is located on chromosome 15q26 [6].

References

[1] Baldwin, C.T. et al (1989) A new epidermal growth factor-like repeat in the human core protein for the large cartilage-specific proteoglycan. J. Biol. Chem. 264: 15747–15750.

[2] Doege, K. et al (1991) Complete coding sequence and deduced primary structure of the human cartilage large aggregating proteoglycan, aggrecan. J. Biol. Chem. 262: 894–902.

[3] Flanney, C.R. et al (1992) Identification of a stromelysin cleavage site within the interglobular repeat of human aggrecan. Evidence for proteolysis at this site in vivo in human articular cartilage. J. Biol. Chem. 267: 1008–1014.

[4] Heinegård, D. (1977) Polydispersity of cartilage proteoglycans. Structural variations with size and buoyant density of the molecules. J. Biol. Chem. 252: 1980–9.

[5] Vilim, V. and Krajickova, J. (1991) Electrophoretic separation of large proteoglycans in large-pore polyacrylamide gradient gels (1.32–10.0% T) and a one-step procedure for simultaneous staining of proteins and proteoglycans. Anal. Biochem. 197: 34–39.

[6] Korenberg, J.R. et al (1993) Assignment of the human aggrecan gene (AGCI) to 15q26 using fluorescent in situ hybridization analysis. Genomics 16:546–551.

Biglycan
PG-S1, PGI, byglycan

Biglycan is a member of a family of structurally related proteoglycans called the small CS/DS proteoglycans. Biglycan is the predominant small proteoglycan of cartilage and aorta and contains two chains of CS/DS. Rotary shadowing electron microscopy shows two extended GAG chains. It is relatively abundant in the mineral compartment of bone, together with decorin.

Molecular structure

Biglycan contains two CS/DS side-chains, but most of the protein consists of 12 repeats of 24 residues. These leucine-rich sequences share marked homology (80% identity) with sequences in other proteoglycans such as decorin and fibromodulin and in other proteins including the serum protein LRG, platelet surface protein GPIb, ribonuclease/angiotensin inhibitor (RAI), chaoptin, toll protein and adenylate cyclase [1,2].

Isolation

Biglycan can be isolated from a variety of tissues including human bone and articular cartilage by guanidine–HCl, density gradient centrifugation, gel filtration, ion-exchange chromatography and octyl-Sepharose chromatography [3,4].

Accession number

P21810; P13247

Primary structure

Ala	A	14	Cys	C	7	Asp	D	23	Glu	E	18

Ala A 14   Cys C 7   Asp D 23   Glu E 18
Phe F 16   Gly G 24   His H 11   Ile I 18
Lys K 22   Leu L 53   Met M 8   Asn N 27
Pro P 24   Gln Q 11   Arg R 18   Ser S 24
Thr T 12   Val V 23   Trp W 4   Tyr Y 11
Mol. wt (calc.) = 41 550   Residues = 368

```
1    MWPLWRLVSL  LALSQALPFE  QRGFWDFTLD  DGPFMMNDEE  ASGADTSGVL
51   DPDSVTPTYS  AMCPFGCHCH  LRVVQCSDLG  LKSVPKEISP  DTTLLDLQNN
101  DISELRKDDF  KGLQHLYALV  LVNNKISKIH  EKAFSPLRNV  QKLYISKNHL
151  VEIPPNLPSS  LVDVRIHDNR  IRKVPKGVFS  GLRNMNCIEM  GGNPLENSGF
201  EPGAFDGLKL  NYLRISEAKL  TGIPKDLPET  LNELHLDHNK  IQAIELEDLL
251  RYSKLYRLGL  GHNQIRMIEN  GSLSFLPTLR  ELHLDNNKLA  RVPSGLPDLK
301  LLQVVYLHSN  NITKVGVNDF  CPMGFGVKRA  YYNGISLFNN  PVPYWEVQPA
351  TFRCVTDRLA  IQFGNYKK
```

Structural and functional sites

Putative signal peptide: 1–19
Propeptide: 20–37
Leucine-rich repeats: 71–94, 95–115, 116–139, 140–163, 164–184, 185–210, 211–230, 231–254, 255–275, 276–301, 302–320, 335–342
Potential glycosaminoglycan attachment sites: 270, 311
Determined glycosaminoglycan attachment sites: 42, 47

Gene structure

The human biglycan gene is a single copy located on chromosome Xq27–qter. It is approximately 8 kb in size and contains eight exons [5].

References

[1] Fisher, L.W. et al (1989) Deduced protein sequence of bone small proteoglycan (biglycan) shows homology with proteoglycan II (decorin) and several non-connective tissue proteins in a variety of species. J. Biol. Chem. 264: 4571–4576.

[2] Roughley, P.J. and White, R.J. (1989) Dermatan sulphate proteoglycans of human articular cartilage. The properties of dermatan sulphate proteoglycans I and II Biochem. J. 262: 823–827.

[3] Fisher, L.W. et al (1987) Purification and partial characterization of small proteoglycans I and II, bone sialoproteins I and II, and osteonectin from the mineral compartment of developing human bone. J. Biol. Chem. 262: 9702–9708.

[4] Choi, H.U. et al (1989) Characterization of the dermatan sulfate proteoglycans, DS-PG1 and DS-PGII, from bovine articular cartilage and skin isolated by octyl-Sepharose chromatography. J. Biol. Chem. 264: 2876–2884.

[5] Fisher, L.W. et al (1991) Human biglycan gene – putative promoter, intron–exon junctions, and chromosomal localization. J. Biol. Chem. 266: 14371–14377.

Bone sialoprotein

BSP, bone sialoprotein II, BSPII

Bone sialoprotein has a restricted tissue distribution, essentially being found only in bone and mineralized connective tissue. In bone, bone sialoprotein is expressed at high levels by osteoblasts at sites of *de novo* bone formation. The molecule binds tightly to hydroxyapatite and appears to form an integral part of the mineralized matrix. Bone sialoprotein promotes cell attachment by binding to the integrin αVβ3.

Molecular structure

Bone sialoprotein is composed of a single polypeptide chain. Its primary sequence reveals a paucity of hydrophobic residues which predicts an open, flexible structure. Bone sialoprotein is isolated from bone matrix as a component of approximate molecular weight 60 000, containing both O- and N-linked oligosaccharides that are rich in sialic acid. The molecule is extensively glycosylated, about half of its serine residues are phosphorylated, and it contains extended sequences of acidic amino acid residues. Glutamic acid, glycine and aspartic acid account for about 30% of all amino acids. The glutamates are clustered, in two cases as eight consecutive residues. Both ends of the molecule, but particularly the COOH-terminus, contain high contents of tyrosine residues, many of which are sulphated. A single RGD cell adhesion site is located near the COOH-terminus in the middle of clusters of sulphated tyrosine residues. The rabbit molecule contains a keratan sulphate chain and is therefore a proteoglycan [1–6].

Isolation

Bone sialoprotein can be isolated from bone by extraction with demineralizing solutions (e.g. 0.5 M EDTA), and purified by molecular sieve and DEAE ion-exchange chromatography [7].

Accession number

P21815

Primary structure

Ala	A	16	Cys	C	1	Asp	D	16	Glu	E	60

Ala A 16   Cys C 1   Asp D 16   Glu E 60
Phe F 7   Gly G 38   His H 5   Ile I 7
Lys K 15   Leu L 11   Met M 3   Asn N 24
Pro P 12   Gln Q 8   Arg R 8   Ser S 25
Thr T 30   Val V 8   Trp W 0   Tyr Y 23
Mol. wt. (calc.) = 34 982   Residues = 317

```
1    MKTALILLSI  LGMACAFSMK  NLHRRVKIED  SEENGVFKYR  PRYYLYKHAY
51   FYPHLKRFPV  QGSSDSSEEN  GDDSSEEEE   EEETSNEGEN  NEESNEDEDS
101  EAENTTLSAT  TLGYGEDATP  GTGYTGLAAI  QLPKKAGDIT  NKATKEKESD
151  EEEEEEEEGN  ENEESEAEVD  ENEQGINGTS  TNSTEAENGN  GSSGGDNGEE
201  GEEESVTGAN  AEGTTETGGQ  GKGTSKTTTS  PNGGFEPTTP  PQVYRTTSPP
251  FGKTTTVEYE  GEYEYTGVNE  YDNGYEIYES  ENGEPRGDNY  RAYEDEYSYF
301  KGQGYDGYDG  QNYYHHQ
```

Structural and functional sites
Signal peptide: 1–16
Poly-Glu: 76–83, 151–158
Potential N-linked glycosylation sites: 104, 177, 182, 190
RGD cell adhesion site: 286–288
Phosphorylation sites: 67, 74, 149, 184
Sulphation sites: 271, 275, 290, 293, 297, 299

Gene structure

The gene appears to be a single copy located on human chromosome 4. The human gene is approximately 15 kb in length [4].

References
[1] Kinne, R.W. and Fisher, L.W. (1987) Keratan sulfate proteoglycan in rabbit compact bone is bone sialoprotein II. J. Biol. Chem. 262: 10206–10211.
[2] Oldberg, A. et al (1988) The primary structure of a cell-binding bone sialoprotein. J. Biol. Chem. 263: 19430–19432.
[3] Oldberg, A. et al (1988) Identification of a bone sialoprotein receptor in osteosarcoma cells. J. Biol. Chem. 263: 19433–19436.
[4] Fisher, L.W. et al (1990) Human bone sialoprotein. Deduced protein sequence and chromosomal localization. J. Biol. Chem. 265: 2347–2351.
[5] Midura, R.J. et al (1990) A rat osteogenic cell line (UMR.106.01) synthesizes a highly sulfated form of bone sialoprotein. J. Biol. Chem. 265: 5285–5291.
[6] Sodek, J. et al (1992) Elucidating the functions of bone sialoprotein and osteopontin in bone formation. In: Chemistry and Biology of Mineralized Tissues, Slavkin, H. and Price, P., eds., Elsevier, Amsterdam, pp. 297–306.
[7] Franzen, A. and Heinegard, D. (1985) Isolation and characterisation of two sialoproteins present only in bone calcified matrix. Biochem. J. 232: 715–724.

Cartilage matrix protein

Cartilage matrix protein is a major component of the extracellular matrix of non-articular cartilage. It is found in tracheal, nasal septal, xiphisternal, auricular and epiphyseal cartilage, but not in articular cartilage or extracts of the intervertebral disc. While cartilage matrix protein has been shown to account for 5% of the wet weight of aged tracheal cartilage, its function is not known, although it has been suggested that it can bind to and bridge type II collagen fibrils.

Molecular structure

Cartilage matrix protein is a 54 kDa protein which occurs in cartilage as a disulphide-bonded multimer. The protein is normally isolated as a homotrimer (molecular weight approximately 148 000). Cartilage matrix protein contains two repeating sequences of 190 amino acids that are homologous to the A-type repeats of von Willebrand factor, separated by a six-cysteine EGF repeat. The CMP1 and CMP2 repeats contain a cysteine at each end which may facilitate intradomain disulphide bonding [1-3].

Isolation

Cartilage matrix protein can be isolated from cartilage by extraction with 4 M guanidine–HCl followed by fractionation on a caesium chloride density gradient under associative conditions. The top of the gradient is then fractionated by size exclusion chromatography on Sephadex G-200 and/or Sepharose 4B [4].

Accession number
P21941

Primary structure

Ala	A	42	Cys	C	14	Asp	D	29	Glu	E	28
Phe	F	22	Gly	G	36	His	H	7	Ile	I	22
Lys	K	33	Leu	L	43	Met	M	8	Asn	N	11
Pro	P	15	Gln	Q	23	Arg	R	26	Ser	S	48
Thr	T	26	Val	V	53	Trp	W	0	Tyr	Y	10

Mol. wt (calc.) = 53 639   Residues = 496

```
1    MRVLSGTSLM   LCSLLLLLQA   LCSPGLAPQS   RGHLCRTRPT   DLVFVVDSSR
51   SVRPVEFEKV   KVFLSQVIES   LDVGPNATRV   GMVNYASTVK   QEFSLRAHVS
101  KAALLQAVRR   IQPLSTGTMT   GLAIQFAITK   AFGDAEGGRS   RSPDISKVVI
151  VVTDGRPQDS   VQDVSARARA   SGVELFAIGV   GSVDKATLRQ   IASEPQDEHV
201  DYVESYSVIE   KLSRKFQEAF   CVVSDLCATG   DHDCEQVCIS   SPGSYTCACH
251  EGFTLNSDGK   TCNVCSGGGG   SSATDLVFLI   DGSKSVRPEN   FELVKKFISQ
301  IVDTLDVSDK   LAQVGLVQYS   SSVRQEFPLG   RFHTKKDIKA   AVRNMSYMEK
351  GTMTGAALKY   LIDNSFTVSS   GARPGAQKVG   IVFTDGRSQD   YINDAAKKAK
401  DLGFKMFAVG   VGNAVEDELR   EIASEPVAEH   YFYTADFKTI   NQIGKKLQKK
451  ICVEEDPCAC   ESLVKFQAKV   EGLLQALTRK   LEAVSKRLAI   LENTVV
```

Structural and functional sites
Signal peptide: 1–22
vWF A repeats: 23–222, 264–453
EGF (6C) repeat: 223–263
Potential N-linked glycosylation site: 76
Potential intramolecular disulphide bonds: 35–221, 265–452

Gene structure

The human cartilage matrix protein gene spans 12 kb and has eight exons. It appears to be a single-copy gene located on chromosome 1 (locus p35) [2,3].

References

[1] Heinegard, D. and Paulsson, M. (1987) Cartilage. Methods Enzymol. 145: 336–363.

[2] Kiss, I., Deak, F. et al (1989) Structure of the gene for cartilage matrix protein, a modular protein of the extracellular matrix. J. Biol. Chem. 264: 8126–8134.

[3] Jenkins, R.N. et al (1990) Structure and chromosomal location of the human gene encoding cartilage matrix protein. J. Biol. Chem. 265: 19624–19631.

[4] Winterbottom, N. et al (1992) Cartilage matrix protein is a component of the collagen fibril of cartilage. Devel. Dynamics 193: 266–276.

Cartilage oligomeric matrix protein

COMP, high molecular weight cartilage matrix glycoprotein

Cartilage oligomeric matrix protein is a glycoprotein found in cartilage which appears at specific times, distinct from type II collagen, during chondrogenesis. Anchorage in the cartilage matrix appears to be dependent on divalent cations. COMP is preferentially localized in the territorial matrix surrounding chondrocytes. It is found in all cartilages and in the vitreous of the eye.

Molecular structure

COMP is a markedly anionic molecule due not only to its high content of aspartic and glutamic acid residues, but also to its substitution with negatively charged sugars. The molecule appears to consist of five disulphide-bonded sub-units (each of approximate molecular weight 100 000) which give a molecular mass of 524 kDa by sedimentation equilibrium centrifugation. Each subunit contains a peripheral globular domain, a flexible strand and an assembly domain where the five arms meet. There are four six-cysteine EGF repeats and seven thrombospondin type 3 repeats. COMP is homologous to the COOH-terminal portion of thrombospondin [1,2].

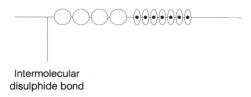

Intermolecular disulphide bond

Isolation

COMP can be isolated from 4 M guanidine–HCl extracts of cartilage by CsCl density gradient centrifugation, Sephadex G-200 chromatography, or DEAE–cellulose chromatography, and can be purified by passing through gelatin–Sepharose and heparin–Sepharose columns [3].

Accession number

S47828

Primary structure (bovine)

Ala A 48	Cys C 46	Asp D 97	Glu E 31
Phe F 26	Gly G 71	His H 15	Ile I 18
Lys K 21	Leu L 37	Met M 9	Asn N 44
Pro P 49	Gln Q 44	Arg R 51	Ser S 34
Thr T 37	Val V 56	Trp T 10	Tyr Y 11

Mol. wt (calc.) = 82 573   Residues = 755

```
  1    MSPTACVLVL   ALAALRATGQ   GQIPLGGDLA   PQMLRELQET   NAALQDVREL
 51    LRHRVKEITF   LKNTVMECDA   CGMQPARTPG   LSVRPVALCA   PGSCFPGVVC
101    TETATGARCG   PCPPGYTGNG   SHCTDVNECN   AHPCFPRVRC   INTSPGFHCE
151    ACPPGFSGPT   HEGVGLTFAK   TNKQVCTDIN   ECETGQHNCV   PNSVCVNTRG
201    SFQCGPCQPG   FVGDQRSGCQ   RRGQHFCPDG   SPSPCHEKAD   CILERDGSRS
251    CVCAVGWAGN   GLLCGRDTDL   DGFPDEKLRC   SERQCRKDNC   VTVPNSGQED
```

```
301   VDRDRIGDAC   DPDADGDGVP   NEQDNCPLVR   NPDQRNSDKD   KWGDACDNCR
351   SQKNDDQKDT   DRDGQGDACD   DDIDGDRIRN   VADNCPRVPN   FDQSDSDGDG
401   VGDACDNCPQ   KDNPDQRDVD   HDFVGDACDS   DQDQDGDGHQ   DSRDNCPTVP
451   NSAQQDSDHD   GKGDACDDDD   DNDGVPDSRD   NCRLVPNPGQ   EDNDRDGVGD
501   ACQGDFDADK   VIDKIDVCPE   NAEVTLTDFR   AFQTVVLDPE   GDAQIDPNWV
551   VLNQGMEIVQ   TMNSDPGLAV   GYTAFNGVDF   EGTFHVNTAT   DDDYAGFIFG
601   YQDSSSFYVV   MWKQMEQTYW   QANPFRAVAE   PGIQLKAVKS   STGPGEQLRN
651   ALWHTGDTAS   QVRLLWKDPR   NVGWKDKTSY   RWFLQHRPQV   GYIRVRFYEG
701   PELVADSNVV   LDTAMRGGRL   GVFCFSQENI   IWANLRYRCN   DTIPEDYERH
751   RLRRA
```

Structural and functional sites

Signal peptide: 1–19
EGF (6C) repeats: 87–126, 127–179, 180–224, 225–267
Thrombospondin type 3 repeats: 295–330, 331–353, 354–389, 390–412, 413–450, 451–486, 487–522
Potential N-glycosylation sites: 119, 142, 730

Gene structure

Transfer blot analysis of bovine articular chondrocyte RNA shows a single 2.5 kb transcript. Gene structure is unknown.

References

[1] Morgelin, M. et al (1992) Electron microscopy of native cartilage oligomeric matrix protein purified from the swarm rat chondrosarcoma reveals a five-armed structure. J. Biol. Chem. 267: 6137–6141

[2] Oldberg, A. et al (1992) COMP (cartilage oligomeric matrix protein) is structurally related to the thrombospondins. J. Biol. Chem. 267: 22346–22350

[3] Hedbom, E. et al (1992) Cartilage matrix proteins. An acidic oligomeric protein (COMP) detected only in cartilage. J. Biol. Chem. 267: 6132–6136.

Collagen type I

Type I collagen is the major fibrillar collagen of bone, tendon and skin; it provides these and many other tissues with tensile strength. Type I collagen forms rope-like structures in tendon, sheet-like structures in skin, and in bone is reinforced with calcium hydroxyapatite.

Molecular structure

Type I collagen is synthesized primarily as a heterotrimeric procollagen comprising two proα1(I) chains and one proα2(I) chain. The procollagen is processed extracellularly by N- and C-proteinases to give a triple-helical molecule that can assemble with a stagger of 234 amino acids into cross-banded fibrils with a 67 nm (D) periodicity. These fibrils are stabilized by intermolecular cross-links derived from specific lysine/hydroxylysine residues in both the non-helical (telopeptide) and helical domains. The fibrillar form of type I collagen interacts with the protein core of the proteoglycan decorin. The triple-helical domain interacts with cells via integrin receptors, principally α2β1, and a tetrapeptide DGEA has been reported to account for this binding. A homopolymer comprising three identical α1(I) chains (designated collagen type I trimer) has also been observed in foetal and diseased tissues but it is not a significant component of normal adult tissues [1-5].

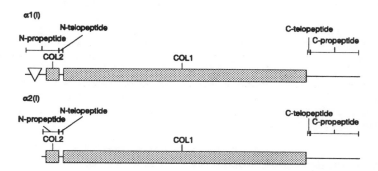

Isolation

Type I collagen can be prepared from foetal skin in its intact, processed form by extraction with 0.5 M acetic acid or in a slightly shorter form by pepsin digestion [6].

Accession number

P02452

Primary structure: α1(I) chain

Sequence conflicts: 59 R to Q
1434 T to S

Ala	A	141	Cys	C	18	Asp	D	66	Glu	E	75
Phe	F	27	Gly	G	391	His	H	9	Ile	I	24
Lys	K	58	Leu	L	48	Met	M	13	Asn	N	28
Pro	P	278	Gln	Q	48	Arg	R	71	Ser	S	60

Thr T 44   Val V 46   Trp W 6   Tyr Y 13
Mol. wt (calc.) = 138 731   Residues = 1464

```
1      MFSFVDLRLL  LLLAATALLT  HGQEEGQVEG  QDEDIPPITC  VQNGLRYHDR
51     DVWKPEPCRI  CVCDNGKVLC  DDVICDETKN  CPGAEVPEGE  CCPVCPDGSE
101    SPTDQETTGV  EGPKGDTGPR  GPRGPAGPPG  RDGIPGQPGL  PGPPGPPGPP
151    GPPGLGGNFA  PQLSYGYDEK  STGGISVPGP  MGPSGPRGLP  GPPGAPGPQG
201    FQGPPGEPGE  PGASGPMGPR  GPPGPPGKNG  DDGEAGKPGR  PGERGPPGPQ
251    GARGLPGTAG  LPGMKGHRGF  SGLDGAKGDA  GPAGPKGEPG  SPGENGAPGQ
301    MGPRGLPGER  GRPGAPGPAG  ARGNDGATGA  AGPPGPTGPA  GPPGFPGAVG
351    AKGEAGPQGP  RGSEGPQGVR  GEPGPPGPAG  AAGPAGNPGA  DGQPGAKGAN
401    GAPGIAGAPG  FPGARGPSGP  QGPGGPPGPK  GNSGEPGAPG  SKGDTGAKGE
451    PGPVGVQGPP  GPAGEEGKRG  ARGEPGPTGL  PGPPGERGGP  GSRGFPGADG
501    VAGPKGPAGE  RGSPGPAGPK  GSPGEAGRPG  EAGLPGAKGL  TGSPGSPGPD
551    GKTGPPGPAG  QDGRPGPPGP  PGARGQAGVM  GFPGPKGAAG  EPGKAGERGV
601    PGPPGAVGPA  GKDGEAGAQG  PPGPAGPAGE  RGEQGPAGSP  GFQGLPGPAG
651    PPGEAGKPGE  QGVPGDLGAP  GPSGARGERG  FPGERGVQGP  PGPAGPRGAN
701    GAPGNDGAKG  DAGAPGAPGS  QGAPGLQGMP  GERGAAGLPG  PKGDRGDAGP
751    KGADGSPGKD  GVRGLTGPIG  PPGPAGAPGD  KGESGPSGPA  GPTGARGAPG
801    DRGEPGPPGP  AGFAGPPGAD  GQPGAKGEPG  DAGAKGDAGP  PGPAGPAGPP
851    GPIGNVGAPG  AKGARGSAGP  PGATGFPGAA  GRVGPPGPSG  NAGPPGPPGP
901    AGKEGGKGPR  GETGPAGRPG  EVGPPGPPGP  AGEKGSPGAD  GPAGAPGTPG
951    PQGIAGQRGV  VGLPGQRGER  GFPGLPGPSG  EPGKQGPSGA  SGERGPPGPM
1001   GPPGLAGPPG  ESGREGAPGA  EGSPGRDGSP  GAKGDRGETG  PAGPPGAPGA
1051   PGAPGPVGPA  GKSGDRGETG  PAGPAGPVGP  AGARGPAGPQ  GPRGDKGETG
1101   EQGDRGIKGH  RGFSGLQGPP  GPPGSPGEQG  PSGASGPAGP  RGPPGSAGAP
1151   GKDGLNGLPG  PIGPPGPRGR  TGDAGPVGPP  GPPGPPGPPG  PPSAGFDFSF
1201   LPQPPQEKAH  DGGRYYRADD  ANVVRDRDLE  VDTTLKSLSQ  QIENIRSPEG
1251   SRKNPARTCR  DLKMCHSDWK  SGEYWIDPNQ  GCNLDAIKVF  CNMETGETCV
1301   YPTQPSVAQK  NWYISKNPKD  KRHVWFGESM  TDGFQFEYGG  QGSDPADVAI
1351   QLTFLRLMST  EASQNITYHC  KNSVAYMDQQ  TGNLKKALLL  KGSNEIEIRA
1401   EGNSRFTYSV  TVDGCTSHTG  AWGKTVIEYK  TTKTSRLPII  DVAPLDVGAP
1451   DQEFGFDVGP  VCFL
```

Structural and functional sites
Signal peptide: 1–22
N-Propeptide: 23–161
von Willebrand factor C repeat 35–103
   COL2 domain: 109–159
N-Telopeptide: 162–178
Helical domain: 179–1192
C-Telopeptide: 1193–1218
C-Propeptide: 1219–1464
Lysine/hydroxylysine cross-linking sites: 170, 265, 1108, 1208
Potential N-linked glycosylation site: 1365
N-Proteinase cleavage site: 161–162
C-Proteinase cleavage site: 1218–1219
DGEA cell adhesion site: 613–616
Mammalian collagenase cleavage site: 953–954

Accession number
P02464; P08123

Primary structure: α2(I) chain

Sequence conflicts:
249	N	to	I
276	T	to	A
333	P	to	V
338	T	to	A
483	V	to	A
549	A	to	D
743	G	to	A
828	V	to	A
831	T	to	P
837	V	to	P
1098	L	to	P
1101	L	to	P

Ala	A	126	Cys	C	9	Asp	D	42	Glu	E	67
Phe	F	23	Gly	G	382	His	H	17	Ile	I	31
Lys	K	50	Leu	L	62	Met	M	10	Asn	N	42
Pro	P	230	Gln	Q	32	Arg	R	72	Ser	S	51
Thr	T	44	Val	V	55	Trp	W	5	Tyr	Y	16

Mol. wt (calc.) = 129 357   Residues = 1366

```
1      MLSFVDTRTL   LLLAVTLCLA   TCQSLQEETV   RKGPAGDRGP   RGERGPPGPP
51     GRDGEDGPTG   PPGPPGPPGP   PGLGGNFAAQ   YDGKGVGLGP   GPMGLMGPRG
101    PPGAAGAPGP   QGFQGPAGEP   GEPGQTGPAG   ARGPAGPPGK   AGEDGHPGKP
151    GRPGERGVVG   PQGARGFPGT   PGLPGFKGIR   GHNGLDGLKG   QPGAPGVKGE
201    PGAPGENGTP   GQTGARGLPG   ERGRVGAPGP   AGARGSDGSV   GPVGPAGPNG
251    SAGPPGFPGA   PGPKGEIGAV   GNAGPTGPAG   PRGEVGLPGL   SGPVGPPGNP
301    GANGLTGAKG   AAGLPGVAGA   PGLPGPRGIP   GPPGAAGTTG   ARGLVGEPGP
351    AGSKGESGNK   GEPGSAGPQG   PPGPSGEEGK   RGPNGEAGSA   GPPGPPGLRG
401    SPGSRGLPGA   DGRAGVMGPP   GSRGASGPAG   VRGPNGDAGR   PGEPGLMGPR
451    GLPGSPGNIG   PAGKEGPVGL   PGIDGRPGPI   GPVGARGEPG   NIGFPGPKGP
501    TGDPGKNGDK   GHAGLAGARG   APGPDGNNGA   QGPPGPQGVQ   GGKGEQGPAG
551    PPGFQGLPGP   SGPAGEVGKP   GERGLHGEFG   LPGPAGPRGE   RGPPGESGAA
601    GPTGPIGSRG   PSGPPGPDGN   KGEPGVVGAV   GTAGPSGPSG   LPGERGAAGI
651    PGGKGEKGEP   GLRGEIGNPG   RDGARGAHGA   VGAPGPAGAT   GDRGEAGAAG
701    PAGPAGPRGS   PGERGEVGPA   GPNGFAGPAG   AAGQPGAKGE   RGGKGPKGEN
751    GVVGPTGPVG   AAGPAGPNGP   PGPAGSRGDG   GPPGMTGFPG   AAGRTGPPGP
801    SGISGPPGPP   GPAGKEGLRG   PRGDQGPVGR   TGEVGAVGPP   GFAGEKGPSG
851    EAGTAGPPGT   PGPQGLLGAP   GILGLPGSRG   ERGLPGVAGA   VGEPGPLGIA
901    GPPGARGPPG   AVGSPGVNGA   PGEAGRDGNP   GNDGPPGRDG   QPGHKGERGY
951    PGNIGPVGAA   GAPGPHGPVG   PAGKHGNRGE   TGPSGPVGPA   GAVGPRGPSG
1001   PQGIRGDKGE   PGEKGPRGLP   GFKGHNGLQG   LPGIAGHHGD   QGAPGSVGPA
1051   GPRGPAGPSG   PAGKDGRTGH   PGTVGPAGIR   GPQGHQGPAG   PPGPPGPLGP
1101   LGVSGGGYDF   GYDGDFYRAD   QPRSAPSLRP   KDYEVDATLK   SLNNQIETLL
1151   TPEGSRKNPA   RTCRDLRLSH   PEWSSGYYWI   DPNQGCTMEA   IKVYCDFPTG
1201   ETCIRAQPEN   IPAKNWYRSS   KDKKHVWLGE   TINAGSQFEY   NVEGVTSKEM
1251   ATQLAFMRLL   ANYASQNITY   HCKNSIAYMD   EETGNLKKAV   ILQGSNDVEL
```

```
1301  VAEGNSRFTY  TVLVDGCSKK  TNEWGKTIIE  YKTNKPSRLP  FLDIAPLDIG
1351  GADHEFFVDI  GPVCFK
```

Structural and functional sites

Signal peptide: 1–22
N-Propeptide: 23–79
  COL2 domain: 33–77
N-Telopeptide: 80–90
Helical domain: 91–1102
C-Telopeptide: 1103–1119
C-Propeptide: 1120–1366
Lysine/hydroxylysine cross-linking sites: 84, 177, 1023
Histidine cross-linking site: 182
Hydroxylysine glycosylation sites: 177, 264
Potential N-linked glycosylation site: 1267
N-Proteinase cleavage site: 79–80
C-Proteinase cleavage site: 1119–1120
Mammalian collagenase cleavage site: 865–866

Gene structure

The proα1(I) and proα2(I) chains are encoded by single genes located on human chromosomes 17 (locus q21.3–22) and 7 (locus q21.3–22), respectively. The proα1(I) gene contains 51 exons and the proα2(I) gene 52 exons. All the exons encoding the triple-helical domain are multiples of 9 bp corresponding to Gly–X–Y triplets (commonly 54 bp). The exon arrangement within the uninterrupted triple-helical domain (exons 7–47 for α1, exons 7–48 for α2(I)) are almost identical to each other and to that for exons 9–50 in the α1(II) gene. In cartilage, the use of a different transcription site within the first intron of the chick proα2(I) gene results in exons 1 and 2 being replaced by a new exon of 96 bp. The resulting transcript contains several open reading frames that are out of frame with the α2(I) coding sequence and therefore encode non-collagenous proteins, one of which appears to be a DNA binding protein. Whether the α1(I) chain of collagen type I trimer is the same or a distinct gene product from that in the heterotrimer is still controversial [7–9].

References

[1] Fietzek, P.P. and Kuhn, K. (1976) The primary structure of collagen. Int. Rev. Connective Tissue Res. 7: 1–60.

[2] Schupp-Byrne, D.E. and Church, R.L. (1982) Embryonic collagen (type I-trimer) α1-chains are genetically distinct from type I collagen α1-chains. Collagen Rel. Res. 2: 481–494.

[3] Kuhn, K. (1987) The classical collagens: types I, II and III. In: Structure and Function of Collagen Types, Mayne, R. and Burgeson, R.E., eds, Academic Press, London, pp. 1–42.

[4] Vuorio, E. and de Crombrugghe, B. (1990) The family of collagen genes. Annu. Rev. Biochem. 59: 837–872.

[5] Staatz, W.D. et al (1991) Identification of a tetrapeptide recognition sequence for the α2β1 integrin in collagen. J. Biol. Chem. 266: 7363–7367.

6 Miller, E.J. and Rhodes, R.K. (1982) Preparation and characterization of the different types of collagen. Methods Enzymol. 82: 33–64.

7 Bennett, V. and Adams, S.L. (1990) Identification of a cartilage-specific promoter within intron 2 of the chick α2(I) collagen gene. J. Biol. Chem. 265: 2223–2230.

8 Sandell, L.J. and Boyd, C.D. (1990) Conserved and divergent sequence and functional elements within collagen genes. In: Extracellular Matrix Genes, Sandell, L.J. and Boyd, C.D., eds, Academic Press, New York, pp. 1–56.

9 Chu, M.L. and Prockop, D.J. (1993) Collagen: Gene Structure. In: Connective Tissue and Its Heritable Disorders, Steinmann, B. and Royce, P., eds, Wiley-Liss, New York, pp. 149–165.

Collagen type II

Type II collagen is the principal collagenous component of cartilage, intervertebral disc and vitreous humour, but is also found in other tissues during development. Its function is to provide tensile strength and confers on cartilage the ability to resist shearing forces. Type II collagen supports chondrocyte adhesion and may influence the differentiated phenotype of these cells.

Molecular structure

Type II collagen is synthesized as a homotrimeric procollagen comprising three identical proα1(II) chains. The procollagen is processed extracellularly by N- and C-proteinases to produce a triple-helical molecule that can assemble with a stagger of 234 amino acids into cross-banded fibrils with 67 nm (D) periodicity. These fibrils are stabilized by intermolecular cross-links derived from specific lysine/hydroxylysine residues in both the non-helical (telopeptide) and helical domains. Different molecular forms of type II collagen arise by alternative splicing of the N-propeptide domain. The COL1 domain of type II collagen interacts with integrin receptors, principally α2β1, and the non-integrin binding protein anchorin CII. The fibrillar form of type II collagen is cross-linked to type IX collagen via lysine/hydroxylysine residues in the N- and C-telopeptides of type II collagen and in the helical COL2 domain of type IX collagen. The fibrillar form of type II collagen also interacts with the protein cores of the proteoglycans fibromodulin and decorin [1-3].

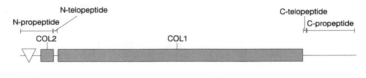

Isolation
Type II collagen can be prepared from young cartilage by guanidine–HCl treatment (to remove proteoglycans) and pepsin digestion [4].

Accession number
P02458

Primary structure: α1(II) chain

Sequence conflicts: 9 S to T
608 G to A
715 G to A
763–766 TSGI to PAGF
988 D to N
1000 A to T
1264 K to R
1281 G to A

Ala	A	132	Cys	C	19	Asp	D	63	Glu	E	79

Ala A 132   Cys C 19   Asp D 63   Glu E 79
Phe F 24   Gly G 405   His H 8   Ile I 35
Lys K 67   Leu L 57   Met M 16   Asn N 32
Pro P 268   Gln Q 61   Arg R 72   Ser S 50
Thr T 44   Val V 38   Trp W 7   Tyr Y 10
Mol. wt (calc.) = 141 788   Residues = 1487

```
1     MIRLGAPQSL  VLLTLLVAAV  LRCQGQDVQE  AGSCVQDGQR  YNDKDVWKPE
51    PCRICVCDTG  TVLCDDIICE  DVKDCLSPEI  PFGECCPICP  TDLATASGQP
101   GPKGQKGEPG  DIKDIVGPKG  PPGPQGPAGE  QGPRGDRGDK  GEKGAPGPRG
151   RDGEPGTLGN  PGPPGPPGPP  GPPGLGGNFA  AQMAGGFDEK  AGGAQLGVMQ
201   GPMGPMGPRG  PPGPAGAPGP  QGFQGNPGEP  GEPGVSGPMG  PRGPPGPPGK
251   PGDDGEAGKP  GKAGERGPPG  PQGARGFPGT  PGLPGVKGHR  GYPGLDGAKG
301   EAGAPGVKGE  SGSPGENGSP  GPMGPRGLPG  ERGRTGPAGA  AGARGNDGQP
351   GPAGPPGPVG  PAGGPGFPGA  PGAKGEAGPT  GARGPEGAQG  PRGEPGTPGS
401   PGPAGASGNP  GTDGIPGAKG  SAGAPGIAGA  PGFPGPRGPP  DPQGATGPLG
451   PKGQTGKPGI  AGFKGEQGPK  GEPGPAGPQG  APGPAGEEGK  RGARGEPGGV
501   GPIGPPGERG  APGNRGFPGQ  DGLAGPKGAP  GERGPSGLAG  PKGANGDPGR
551   PGEPGLPGAR  GLTGRPGDAG  PQGKVGPSGA  PGEDGRPGPP  GPQGARGQPG
601   VMGFPGPKGA  NGEPGKAGEK  GLPGAPGLRG  LPGKDGETGA  EGPPGPAGPA
651   GERGEQGAPG  PSGFQGLPGP  PGPPGEGGKP  GDQGVPGEAG  APGLVGPRGE
701   RGFPGERGSP  GAQGLQGPRG  LPGTPGTDGP  KGASGPAGPP  GAQGPPGLQG
751   MPGERGAAGI  AGPKGDRGDV  GEKGPEGAPG  KDGGRGLTGP  IGPPGPAGAN
801   GEKGEVGPPG  PAGSAGARGA  PGERGETGPP  GTSGIAGPPG  ADGQPGAKGE
851   QGEAGQKGDA  GAPGPQGPSG  APGPQGPTGV  TGPKGARGAQ  GPPGATGFPG
901   AAGRVGPPGS  NGNPGPPGPP  GPSGKDGPKG  ARGDSGPPGR  AGEPGLQGPA
951   GPPGEKGEPG  DDGPSGAEGP  PGPQGLAGQR  GIVGLPGQRG  ERGFPGLPGP
1001  SGEPGQQGAP  GASGDRGPPG  PVGPPGLTGP  AGEPGREGSP  GADGPPGRDG
1051  AAGVKGDRGE  TGAVGAPGAP  GPPGSPGPAG  PTGKQGDRGE  AGAQGPMGPS
1101  GPAGARGIQG  PQGPRGDKGE  AGEPGERGLK  GHRGFTGLQG  LPGPPGPSGD
1151  QGASGPAGPS  GPRGPPGPVG  PSGKDGANGI  PGPIGPPGPR  GRSGETGPAG
1201  PPGNPGPPGP  PGPPGPGIDM  SAFAGLGPRE  KGPDPLQYMR  ADQAAGGLRQ
1251  HDAEVDATLK  SLNNQIESIR  SPEGSRKNPA  RTCRDLKLCH  PEWKSGDYWI
1301  DPNQGCTLDA  MKVFCNMETG  ETCVYPNPAN  VPKKNWWSSK  SKEKKHIWFG
1351  ETINGGFHFS  YGDDNLAPNT  ANVQMTFLRL  LSTEGSQNIT  YHCKNSIAYL
1401  DEAAGNLKKA  LLIQGSNDVE  IRAEGNSRFT  YTALKDGCTK  HTGKWGKTVI
1451  EYRSQKTSRL  PIIDIAPMDI  GGPEQEFGVD  IGPVCFL
```

Structural and functional sites
Signal peptide: 1–25
N-Propeptide: 26–181 (amino acid 29 is Q if unspliced, R if spliced)
von Willebrand factor C repeats (alternatively spliced domain): 29–97
   COL2 domain: 98–179
N-Telopeptide: 182–200
Helical domain: 201–1214
C-Telopeptide: 1215–1241
C-Propeptide (chondrocalcin): 1242–1487

Lysine/hydroxylysine cross-linking sites: 190, 287, 1130, 1231
Potential N-linked glycosylation sites: 317, 1388
Interchain disulphide bond residues: 1283, 1289
N-Proteinase cleavage site: 181–182
C-Proteinase cleavage site: 1241–1242
Mammalian collagenase cleavage site: 975–976
Stromelysin cleavage sites: 194–195, 198–199

Gene structure

Type II collagen is encoded by a single gene found on human chromosome 12 at locus q13.11–12. The gene spans approximately 30 kb and contains 54 exons, a number of which are multiples of 9 bp corresponding to Gly–X–Y triplets (commonly 54 bp). Exon 2, the sequence of which is conserved in types I and III collagen and which codes for part of the N-propeptide, can be alternatively spliced [5–7].

References

[1] Kuhn, K. (1987) The classical collagens: Types I, II and III. In: Structure and Function of Collagen Types, Mayne, R. and Burgeson, R.E., eds, Academic Press, London, pp. 1–42.

[2] Ryan, M.C. and Sandell, L.J. (1990) Differential expression of a cysteine-rich domain in the amino-terminal propeptide of type II (cartilage) procollagen by alternative splicing of mRNA. J. Biol. Chem. 265: 10334–10339.

[3] Wu, J.-J. et al (1991) Sites of stromelysin cleavage in collagen types II, IX, X and XI of cartilage. J. Biol. Chem. 266: 5625–5628.

[4] Miller, E.J. and Rhodes, R.K. (1982) Preparation and characterization of the different types of collagen. Methods Enzymol. 82: 33–64.

[5] Ala-Kokko, L. and Prockop, D.J. (1990) Completion of the intron–exon structure of the gene for human type II procollagen (COL2A1): variations in the nucleotide sequences of the allelles from three chromosomes. Genomics 8: 454–460.

[6] Sandell, L.J. and Boyd, C.D. (1990) Conserved and divergent sequence and functional elements within collagen genes. In: Extracellular Matrix Genes, Sandell, L.J. and Boyd, C.D., eds, Academic Press, New York, pp. 1–56.

[7] Chu, M.L. and Prockop, D.J. (1993) Collagen: Gene structure. In: Connective Tissue and Its Heritable Disorders, Steinmann, B. and Royce, P., eds, Wiley-Liss, New York, pp. 149–165.

Collagen type III

Type III collagen is a major fibrillar collagen in skin and vascular tissues. These tissues also contain collagen type I. Type III collagen fibres are thin and result in a more compliant tissue.

Molecular structure

Type III collagen is synthesized as a homotrimeric procollagen comprising three identical proα1(III) chains. The procollagen is processed extracellularly by N- and C-proteinases to produce a triple-helical molecule in which the three α-chains are linked by disulphide bonds at the COOH-terminal end of the triple helix. The molecules assemble with a stagger of 234 amino acids into cross-banded fibrils with a 67 nm (D) periodicity. N-Proteinase cleavage is often incomplete giving rise to a partially processed collagen (pN-collagen) which affects fibril formation. The fibrils are stabilized by intermolecular cross-links derived from specific lysine/hydroxylysine residues in both the non-helical (telopeptide) and helical domains [1-2].

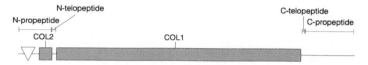

Isolation

Type III collagen can be prepared from foetal skin as pN-collagen and processed forms by NaCl extraction [3], or as the shortened triple-helical α1(III) chain by pepsin digestion [4].

Accession number
P02461

Primary structure: α1(III) chain

Sequence conflicts: 158   G to V
163   G to GG
226–228   missing
241   E to D
278   T to A
293–295 NGA   to DGS
472   E to D
488–490 PGF   to LGS
614   T to Y
635   P to R
664   D to E
676   D to N
896   S to A
980   S to A
985–989 ANGLS   to PSGQN
1019 D to Y

1097	T to P	
1153–1154	TS to AT	
1156	H to P	
1156	H to S	
1184	P to S	
1203	A to P	
1210	G to A	
1241	V to A	
1357	L to P	

Ala	A	115	Cys	C	22	Asp	D	55	Glu	E	74
Phe	F	23	Gly	G	413	His	H	14	Ile	I	36
Lys	K	62	Leu	L	48	Met	M	17	Asn	N	41
Pro	P	281	Gln	Q	43	Arg	R	60	Ser	S	73
Thr	T	31	Val	V	36	Trp	W	7	Tyr	Y	15

Mol. wt (calc.) = 138 403   Residues = 1466

```
1     MMSFVQKGSW  LLLALLHPTI  ILAQQEAVEG  GCSHLGQSYA  DRDVWKPEPC
51    QICVCDSGSV  LCDDIICDDQ  ELDCPNPEIP  FGECCAVCPQ  PPTAPTRPPN
101   GQGPQGPKGD  PGPPGIPGRN  GDPGIPGQPG  SPGSPGPPGI  CESCPTGPQN
151   YSPQYDSYDV  KSGVAVGGLA  GYPGPAGPPG  PPGPPGTSGH  PGSPGSPGYQ
201   GPPGEPGQAG  PSGPPGPPGA  IGPSGPAGKD  GESGRPGRPG  ERGLPGPPGI
251   KGPAGIPGFP  GMKGHRGFDG  RNGEKGETGA  PGLKGENGLP  GENGAPGPMG
301   PRGAPGERGR  PGLPGAAGAR  GNDGARGSDG  QPGPPGPPGT  AGFPGSPGAK
351   GEVGPAGSPG  SNGAPGQRGE  PGPQGHAGAQ  GPPGPPGING  SPGGKGEMGP
401   AGIPGAPGLM  GARGPPGPAG  ANGAPGLRGG  AGEPGKNGAK  GEPGPRGERG
451   EAGIPGVPGA  KGEDGKDGSP  GEPGANGLPG  AAGERGAPGF  RGPAGPNGIP
501   GEKGPAGERG  APGPAGPRGA  AGEPGRDGVP  GGPGMRGMPG  SPGGPGSDGK
551   PGPPGSQGES  GRPGPPGPSG  PRGQPGVMGF  PGPKGNDGAP  GKNGERGGPG
601   GPGPQGPPGK  NGETGPQGPP  GPTGPGGDKG  DTGPPGPQGL  QGLPGTGGPP
651   GENGKPGEPG  PKGDAGAPGA  PGGKGDAGAP  GERGPPGLAG  APGLRGGAGP
701   PGPEGGKGAA  GPPGPPGAAG  TPGLQGMPGE  RGGLGSPGPK  GDKGEPGGPG
751   ADGVPGKDGP  RGPTGPIGPP  GPAGQPGDKG  EGGAPGLPGI  AGPRGSPGER
801   GETGPPGPAG  FPGAPGQNGE  PGGKGERGAP  GEKGEGGPPG  VAGPPGGSGP
851   AGPPGPQGVK  GERGSPGGPG  AAGFPGARGL  PGPPGSNGNP  GPPGPSGSPG
901   KDGPPGPAGN  TGAPGSPGVS  GPKGDAGQPG  EKGSPGAQGP  PGAPGPLGIA
951   GITGARGLAG  PPGMPGPRGS  PGPQGVKGES  GKPGANGLSG  ERGPPGPQGL
1001  PGLAGTAGEP  GRDGNPGSDG  LPGRDGSPGG  KGDRGENGSP  GAPGAPGHPG
1051  PPGPVGPAGK  SGDRGESGPA  GPAGAPGPAG  SRGAPGPQGP  RGDKGETGER
1101  GAAGIKGHRG  FPGNPGAPGS  PGPAGQQGAI  GSPGPAGPRG  PVGPSGPPGK
1151  DGTSGHPGPI  GPPGPRGNRG  ERGSEGSPGH  PGQPGPPGPP  GAPGPCCGGV
1201  GAAAIAGIGG  EKAGGFAPYY  GDEPMDFKIN  TDEIMTSLKS  VNGQIESLIS
1251  PDGSRKNPAR  NCRDLKFCHP  ELKSGEYWVD  PNQGCKLDAI  KVFCNMETGE
1301  TCISANPLNV  PRKHWWTDSS  AEKKHVWFGE  SMDGGFQFSY  GNPELPEDVL
1351  DVQLAFLRLL  SSRASQNITY  HCKNSIAYMD  QASGNVKKAL  KLMGSNEGEF
1401  KAEGNSKFTY  TVLEDGCTKH  TGEWSKTVFE  YRTRKAVRLP  IVDIAPYDIG
1451  GPDQEFGVDV  GPVCFL
```

Structural and functional sites
Signal peptide: 1–23
N-Propeptide: 24–148
von Willebrand factor C repeat: 27–96
    COL2 domain: 103–141
N-Telopeptide: 149–167
Helical domain: 168–1196
C-Telopeptide: 1197–1205
C-Propeptide: 1206–1466
Interchain disulphide bond residues: 141, 144, 1196, 1197
Tyrosine sulphation sites: 151, 155, 158
Lysine/hydroxylysine cross-linking sites: 161, 263, 1106
Hydroxylysine glycosylation site: 263
Potential N-linked glycosylation site: 1367
N-Proteinase cleavage site: 148–149
C-Proteinase cleavage site: 1205–1206
Mammalian collagenase cleavage site: 951–952

Gene structure

The proα1(III) collagen chain is encoded by a single gene found on human chromosome 2 at locus q24.3–31. The gene probably contains 52 exons and, although the human gene has not been fully analysed, to date all the exons encoding the triple-helical domain are multiples of 9 bp corresponding to Gly–X–Y triplets (commonly 54 bp). The gene will exceed 38 kb [5-7].

References
[1] Jukkola, A. et al (1986) Incorporation of sulfate into type III procollagen by cultured human fibroblasts. Identification of tyrosine-O-sulfate. Eur. J. Biochem. 154: 219–224.
[2] Kuhn, K. (1987) The classical collagens: types I, II and III. In: Structure and Function of Collagen Types, Mayne, R. and Burgeson, R.E., eds, Academic Press, London, pp. 1–42.
[3] Byers, P.H. et al (1974) Preparation of type III procollagen and collagen from rat skin. Biochemistry 13: 5243–5248.
[4] Miller, E.J. and Rhodes, R.K. (1982) Preparation and characterization of the different types of collagen. Methods Enzymol. 82: 33–64.
[5] Myers, J.C. and Dion, A.S. (1990) Types III and V procollagens: Homology in genetic organization and diversity in structure. In: Extracellular Matrix Genes, Sandell, L.J. and Boyd, C.D., eds, Academic Press, New York, pp. 57–78.
[6] Sandell, L.J. and Boyd, C.D. (1990) Conserved and divergent sequence and functional elements within collagen genes. In: Extracellular Matrix Genes, Sandell, L.J. and Boyd, C.D., eds, Academic Press, New York, pp. 1–56.
[7] Chu, M.L. and Prockop, D.J. (1993) Collagen: Gene structure. In: Connective Tissue and Its Heritable Disorders, Steinmann, B. and Royce, P., eds, Wiley-Liss, New York, pp. 149–165.

Collagen type IV

Type IV collagen is found exclusively in basement membranes where it provides the major structural support for this matrix. The molecule contains a long (approximately 350 nm) triple-helical domain containing approximately 20 short interruptions that are thought to introduce flexibility into the helix. Type IV collagen self-assembles into a meshwork by the antiparallel interaction and extensive disulphide bonding of four molecules at their NH_2-termini to form the 7S domain, by the interaction of two molecules at their COOH-terminal non-collagenous (NC1) domains, and by lateral aggregations. The assembled type IV collagen meshwork provides a scaffold for the assembly of other basement membrane components through specific interactions with laminin, entactin/nidogen and heparan sulphate proteoglycan. In addition, this meshwork endows the basement membrane with a size-selective filtration property.

Molecular structure

Type IV collagen molecules are composed of three α chains selected from the translated products of six genetically distinct genes (the COL4A1–COL4A6 genes giving rise to the α1(IV)–α6(IV) chains, respectively). The most abundant form of type IV collagen has the composition $[α1(IV)]_2α2(IV)$. The composition of molecules containing the less abundant α3(IV)–α6(IV) chains has yet to be determined, although the chains are expressed in a tissue-specific manner. Cysteine residues in the NC1 domains of all chains sequenced to date are conserved and participate in both intrachain and intermolecular disulphide bonding. The trimeric, disulphide-bonded CB3 cyanogen bromide fragment of type IV collagen has been shown to support cell adhesion via the integrins α1β1 and α2β1 [1-5].

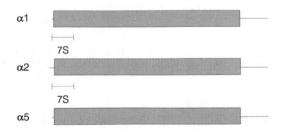

Isolation

Type IV collagen can be isolated from the Engelbreth–Holm–Swarm (EHS) tumour grown in lathyritic C57BL/6 mice by salt (3.4MNaCl) and acid (0.5M acetic acid) exraction, followed by extraction of the insoluble residue with 2m guanidine-HCl. Guanidine-HCl-soluble material is then further purified by DEAE-cellulose ion exchange chromatography [6].

Accession number

P02462

Primary structure: α1(IV) chain

Sequence conflicts: 237–238 SG to KE
241 G to K
719 N to D
837 D to Y
842 K to P
904 E to Q
914 S to W

Ala	A	59	Cys	C	20	Asp	D	58	Glu	E	69
Phe	F	46	Gly	G	478	His	H	16	Ile	I	58
Lys	K	94	Leu	L	92	Met	M	31	Asn	N	16
Pro	P	326	Gln	Q	73	Arg	R	45	Ser	S	70
Thr	T	43	Val	V	51	Trp	W	6	Tyr	Y	18

Mol. wt (calc.) = 160 423   Residues = 1669

```
1      MGPRLSVWLL  LLPAALLLHE  EHSRAAAKGG  CAGSGCGKCD  CHGVKGQKGE
51     RGLPGLQGVI  GFPGMQGPEG  PQGPPGQKGD  TGEPGLPGTK  GTRGPPGASG
101    YPGNPGLPGI  PGQDGPPGPP  GIPGCNGTKG  ERGPLGPPGL  PGFAGNPGPP
151    GLPGMKGDPG  EILGHVPGML  LKGERGFPGI  PGTPGPPGLP  GLQGPVGPPG
201    FTGPPGPPGP  PGPPGEKGQM  GLSFQGPKGD  KGDQGVSGPP  GVPGQAQVQE
251    KGDFATKGEK  GQKGEPGFQG  MPGVGEKGEP  GKPGPRGKPG  KDGDKGEKGS
301    PGFPGEPGYP  GLIGRQGPAG  EKGEAGPPGP  PGIVIGTGPL  GEKGERGYPG
351    TPGPRGEPGP  KGFPGLPGQP  GPPGLPVPGQ  AGAPGFPGER  GEKGDRGFPG
401    TSLPGPSGRD  GLPGPPGSPG  PPGQPGYTNG  IVECQPGPPG  DQGPPGIPGQ
451    PGFIGEIGEK  GQKGESCLIC  DIDGYRGPPG  PQGPPGEIGF  PGQPGAKGDR
501    GLPGRDGVAG  VPGPQGTPGL  IGQPGAKGEP  GEFYFDLRLK  GDKGDPGFFG
551    QPGMPGRAGS  PGRDGHPGLP  GPKGSPGSVG  LKGERGPPGG  VGFPGSRGDT
601    GPPGPPGYGP  AGPIGDKGQA  GFPGGPGSPG  LPGPKGEPGK  IVPLPGPPGA
651    EGLPGSPGFP  GPQGDRGFPG  TPGRPGLPGE  KGAVGQPGIG  FPGPPGPKGV
701    DGLPGDMGPP  GTPGRPGFNG  LPGNPGVQGQ  KGEPGVGLPG  LKGLPGLPGI
751    PGTPGEKGSI  GVPGVPGEHG  AIGPPGLQGI  RGEPGPPGLP  GSVGSPGVPG
801    IGPPGARGPP  GGQGPPGLSG  PPGIKGEKGF  PGFPGLDMPG  PKGDKGAQGL
851    PGITGQSGLP  GLPGQQGAPG  IPGFPGSKGE  MGVMGTPGQP  GSPGPVGAPG
901    LPGEKGDHGF  PGSSGPRGDP  GLKGDKGDVG  LPGKPGSMDK  VDMGSMKGQK
951    GDQGEKGQIG  PIGEKGSRGD  PGTPGVPGKD  GQAGQPGQPG  PKGDPGIKGT
1001   PGAPGLPGPP  GKVGGMGLPG  TPGEKGVPGI  PGPQGSPGLP  GDKGAKGEKG
1051   QAGPPGIGIP  GLRGEKGDQG  IAGFPGSPGE  KGEKGSIGIP  GMPGSPGLKG
1101   SPGSVGYPGS  PGLPGEKGDK  GLPGLDGIPG  VKGEAGLPGT  PGPTGPAGQK
1151   GEPGSDGIPG  SAGEKGEPGL  PGRGFPGFPG  AKGDKGSKGE  VGFPGLAGSP
1201   GIPGSKGEQG  FMGPPGPQGQ  PGLPGSPGHA  TEGPKGDRGP  QGQPGLPGLP
1251   GPMGPPGLPG  IDGVKGDKGN  PGWPGAPGVP  GPKGDPGFQG  MPGIGGSPGI
1301   TGSKGDMGPP  GVPGFQGPKG  LPGLQGIKGD  QGDQGVPGAK  GLPGPPGPPG
1351   PYDIIKGQPG  LPGPEGPPGL  KGLQGLPGPK  GQQGVTGLVG  IPGPPGIPGF
1401   DGAPGQKGEM  GPAGPTGPRG  FPGPPGPDGL  PGSMGPPGTP  SVDHGFLVTR
1451   HSQTIDDPQC  PSGTKILYHG  YSLLYVQGNE  RAHGQDLGTA  GSCLRKFSTM
1501   PFLFCNINNV  CNFASRNDYS  YWLSTPEPMP  MSMAPITGEN  IRPFISRCAV
1551   CEAPAMVMAV  HSQTIQIPPC  PSGWSSLWIG  YSFVMHTSAG  AEGSGQALAS
1601   PGSCLEEFRS  APFIECHGRG  TCNYYANAYS  FWLATIERSE  MFKKPTPSTL
1651   KAGELRTHVS  RCQVCMRRT
```

Structural and functional sites
Signal peptide: 1–27
NH2-terminal non-collagenous domain: 28–42
7S domain: 28–172
Triple-helical domain: 43–1440
COOH-terminal non-collagenous (NC1) domain: 1441–1669
Potential N-linked glycosylation site: 126
Proposed heparin/cell-binding sites: 531–543, 1263–1277

Accession number
P08572

Primary structure: α2(IV) chain

Sequence conflicts: 471   R to P
                683   A to G
            1575   M to I

Ala	A	83	Cys	C	21	Asp	D	82	Glu	E	60

Ala A 83   Cys C 21   Asp D 82   Glu E 60
Phe F 57   Gly G 472   His H 19   Ile I 65
Lys K 82   Leu L 102   Met M 24   Asn N 18
Pro P 286   Gln Q 62   Arg R 78   Ser S 64
Thr T 51   Val V 49   Trp W 9   Tyr Y 28
Mol. wt (calc.) = 167347   Residues = 1712

```
   1    MGRDQRAVAG   PALRRWLLLG   TVTVGFLAQS   VLAGVKKFDV   PCGGRDCSGG
  51    CQCYPEKGGR   GQPGPVGPQG   YNGPPGLQGF   PGLQGRKGDK   GERGAPGVTG
 101    PKGDVGARGV   SGFPGADGIP   GHPGQGGPRG   RPGYDGCNGT   QGDSGPQGPP
 151    GSEGFTGPPG   PQGPKGQKGE   PYALPKEERD   RYRGEPGEPG   LVGFQGPPGR
 201    PGHVGQMGPV   GAPGRPGPPG   PPGPKGQQGN   RGLGFYGVKG   EKGDVGQPGP
 251    NGIPSDTLHP   IIAPTGVTFH   PDQYKGEKGS   EGEPGIRGIS   LKGEEGIMGF
 301    PGLRGYPGLS   GEKGSPGQKG   SRGLDGYQGP   DGPRGPKGEA   GDPGPPGLPA
 351    YSPHPSLAKG   ARGDPGFPGA   QGEPGSQGEP   GDPGLPGPPG   LSIGDGDQRR
 401    GLPGEMGPKG   FIGDPGIPAL   YGGPPGPDGK   RGPPGPPGLP   GPPGPDGFLF
 451    GLKGAKGRAG   FPGLPGSPGA   RGPKGWKGDA   GECRCTEGDE   AIKGLPGLPG
 501    PKGFAGINGE   PGRKGDKGDP   GQHGLPGFPG   LKGVPGNIGA   PGPKGAKGDS
 551    RTITTKGERG   QPGVPGVPGM   KGDDGSPGRD   GLDGFPGLPG   PPGDGIKGPP
 601    GDPGYPGIPG   TKGTPGEMGP   PGLGLPGLKG   QRGFPGDAGL   PGPPGFLGPP
 651    GPAGTPGQID   CDTDVKRAVG   GDRQEAIQPG   CIAGPKGLPG   LPGPPGPTGA
 701    KGLRGIPGFA   GADGGPGPRG   LPGDAGREGF   PGPPGFIGPR   GSKGAVGLPG
 751    PDGSPGPIGL   PGPDGPPGER   GLPGEVLGAQ   PGPRGDAGVP   GQPGLKGLPG
 801    DRGPPGFRGS   QGMPGMPGLK   GQPGLPGPSG   QPGLYGPPGL   HGFPGAPGQE
 851    GPLGLPGIPG   REGLPGDRGD   PGDTGAPGPV   GMKGLSGDRG   DAGFTGEQGH
 901    PGSPGFKGID   GMPGTPGLKG   DRGSPGMDGF   QGMPGLKGRP   GFPGSKGEAG
 951    FFGIPGLKGL   AGEPGFKGSR   GDPGPPGPPP   VILPGMKDIK   GEKGDEGPMG
1001    LKGYLGAKGI   QGMPGIPGLS   GIPGLPGRPG   HIKGVKGDIG   VPGIPGLPGF
1051    PGVAGPPGIT   GFPGFIGSRG   DKGAPGRAGL   YGEIGATGDF   GDIGDTINLP
1101    GRPGLKGERG   TTGIPGLKGF   FGEKGTEGDI   GFPGITGVTG   VQGPPGLKGQ
1151    TGFPGLTGPP   GSQGELGRIG   LPGGKGDDGW   PGAPGLPGFP   GLRGIRGLHG
1201    LPGTKGFPGS   PGSDIHGDPG   FPGPPGERGD   PGEANTLPGP   VGVPGQKGDQ
1251    GAPGERGPPG   SPGLQGFPGI   TPPSNISGAP   GDKGAPGIFG   LKGYRGPPGP
```

```
1301  PGSAALPGSK    GDTGNPGAPG    TPGTKGWAGD    SGPQGRPGVF    GLPGEKGPRG
1351  EQGFMGNTGP    TGAVGDRGPK    GPKGDPGFPG    APGTVGAPGI    AGIPQKIAIQ
1401  PGTVGPQGRR    GPPGAPGEIG    PQGPPGEPGF    RGAPGKAGPQ    GRGGVSAVPG
1451  FRGDEGPIGH    QGPIGQEGAP    GRPGSPGLPG    MPGRSVSIGY    LLVKHSQTDQ
1501  EPMCPVGMNK    LWSGYSLLYF    EGQEKAHNQD    LGLAGSCLAR    FSTMPFLYCN
1551  PGDVCYYASR    NDKSYWLSTT    APLPMMPVAE    DEIKPYISRC    SVCEAPAIAI
1601  AVHSQDVSIP    HCPAGWRSLW    IGYSFLMHTA    AGDEGGGQSL    VSPGSCLEDF
1651  RATPFIECNG    GRGTCHYYAN    KYSFWLTTIP    EQSFQGSPSA    DTLKAGLIRT
1701  HISRCQVCMK    NL
```

Structural and functional sites

Signal peptide: 1–25
NH$_2$-terminal non-collagenous domain: 28–57
7S domain: 26–183
Triple-helical domain: 58–1484
COOH-terminal non-collagenous (NC1) domain: 1485–1712
Potential N-linked glycosylation site: 138

Primary structure: α3(IV) chain

Only limited sequence information is available from the COOH-terminal half of the human and bovine chain (Accession: A39024, A39474, A39786, C39419, S17802, S20672) [7,8].

Primary structure: α4(IV) chain

Only limited sequence information is available from the COOH-terminal half of the bovine chain (Accession: D39419, S18804, S20673, S20834) [9].

Accession number
S19029

Primary structure: α5(IV) chain

Ala A 46	Cys C 20	Asp D 54	Glu E 61
Phe F 41	Gly G 477	His H 13	Ile I 69
Lys K 80	Leu L 113	Met M 26	Asn N 32
Pro P 388	Gln Q 74	Arg R 39	Ser S 65
Thr T 38	Val V 31	Trp W 5	Tyr Y 13

Mol. wt (calc.) = 160 998   Residues = 1685

```
1    MKLRGVSLAA    GLFLLALSLW    GQPAEAAACY    GCSPGSKCDC    SGIKGEKGER
51   GFPGLEGHPG    LPGFPGPEGP    PGPRGQKGDD    GIPGPPGPKG    IRGPPGLPGF
101  PGTPGLPGMP    GHDGAPGPQG    IPGCNGTKGE    RGFPGSPGFP    GLQGPPGPPG
151  IPGMKGEPGS    IIMSSLPGPK    GNPGYPGPPG    IQGLPGPTGI    PGPIGPPGPP
201  GLMGPPGPPG    LPGPKGNMGL    NFQGPKGEKG    EQGLQGPPGP    PGQISEQKRP
251  IDVEFQKGDQ    GLPGDRGPPG    PPGIRGPPGP    PGGEKGEKGE    QGEPGKRGKP
301  GKDGENGQPG    IPGLPGDPGY    PGEPGRDGEK    GQKGDTGPPG    PPGLVIPRPG
351  TGITIGEKGN    IGLPGLPGEK    GERGFPGIQG    PPGLPGPPGA    AVMGPPGPPG
401  FPGERGQKGD    EGPPGISIPG    PPGLDGQPGA    PGLPGPPGPG    SPHIPPSDEI
```

43

```
 451  CEPGPPGPPG  SPGDKGLQGE  QGVKGDKGDT  CFNCIGTGIS  GPPGQPGLPG
 501  LPGPPGSLGF  PGQKGEKGQA  GATGPKGLPG  IPGAPGAPGF  PGSKGEPGDI
 551  LTFPGMKGDK  GELGSPGAPG  LPGLPGTPGQ  DGLPGLPGPK  GEPGGITFKG
 601  ERGPPGNPGL  PGLPGNIGPM  GPPGLALQGP  VGEKGIQGVA  GNPGQPGIPG
 651  PKGDPGQTIT  QPGKPGFRGN  PGRDGDVGLP  GDPGLPGQPG  LPGIPGSKGE
 701  PGIPGIGLPG  PPGPKGFPGI  PGPPGAPGTP  GRIGLEGPPG  PPGFPGPKGA
 751  PGFALPGPPG  PPGLPGFKGA  LGPKGDRGFP  GPPGPPGRTG  LDGLPGPKGD
 801  VGPNGQPGPM  GPPGLPGIGV  QGPPGPPGIP  GPIGQPGLHG  IPGEKGDPGP
 851  PGLDVPGPPG  ERGSPGIPGA  PGPIGPPGSP  GLPGKAGRSG  FPGTKGEMGM
 901  MGPPGPPGPL  GIPGRSGVPG  LKGDDGLQGQ  PGLPGPTGEK  GSKGEPGLPG
 951  PPGPMDPNLL  GSKGEKGEPG  LPGIPGVSGP  KGYQGLPGDP  GQPGLSGQPG
1001  LPGPPGPKGN  PGLPGQPGLI  GPPGLKGTIG  DMGFPGPQGV  EGPPGPSGVP
1051  GQPGSPGLPG  QKGDKGDPGI  SSIGLPGLPG  PKGEPGLPGY  PGNPGIKGSV
1101  GDPGLPGLPG  TPGAKGQPGL  PGFPGTPGPP  GPKGISGPPG  NPGLPGEPGP
1151  VGGGGHPGQP  GPPGEKGKPG  QDGIPGPAGQ  KGEPGQPGFG  NPGPPGLPGL
1201  SGQKGDGGLP  GIPGNPGLPG  PKGEPGFHGF  PGVQGPPGPP  GSPGPALEGP
1251  KGNPGPQGPP  GRPGLPGPEG  PPGLPGNGGI  KGEKGNPGQP  GLPGLPGLKG
1301  DQGPPGLQGN  PGRPGLNGMK  GDPGLPGVPG  FPGMKGPSGV  PGSAGPEGEP
1351  GLIGPPGPPG  LPGPSGQSII  IKGDAGPPGI  PGQPGLKGLP  GPQGPQGLPG
1401  PTGPPGDPGR  NGLPGFDGAG  GRKGDPGLPG  QPGTRGLDGP  PGPDGLQGPP
1451  GPPGTSSVAH  GFLITRHSQT  TDAPQCPQGT  LQVYEGFSLL  YVQGNKRAHG
1501  QDLGTAGSCL  RRFSTMPFMF  CNINNVCNFA  SRNDYSYWLS  TPEPMPMSMQ
1551  PLKGQSIQPF  ISRCAVCEAP  AVVIAVHSQT  IQIPHCPQGW  DSLWIGYSFM
1601  MHTSAGAEGS  GQALASPGSC  LEEFRSAPFI  ECHGRGTCNY  YANSYSFWLA
1651  TVDVSDMFSK  PQSETLKAGD  LRTRISRCQV  CMKRT
```

Structural and functional sites

Signal peptide: 1–26
NH2-terminal non-collagenous domain: 27–41
Triple-helical domain: 42–1456
COOH-terminal non-collagenous (NC1) domain: 1457–1685
Potential N-linked glycosylation site: 125

Primary Structure: α6(IV) chain

Only limited sequence information is available from the NH2-terminal one-third of the human chain (Accession: L22763) [10].

Gene structure

The α1(IV) gene comprises 52 exons dispersed over more than 100 kb of genomic DNA. Exon sizes range from 27 to 192 bp. The first intron is at least 18 kb in length. The exon structure of the α2(IV) gene is not conserved with respect to the α1(IV) gene. The promoter for the α1(IV) chain is arranged head-to-head with the promoter for the α2(IV) gene and the two gene products are transcribed from this common bidirectional promoter in opposite directions. The transcription start sites of the α1 and α2 chains are less than 150 bp apart. Both genes are localized to chromosome 13q34. Similar head-to-head organisations for the genes encoding α3(IV) and α4(IV), and for α5(IV) and α6(IV) are apparent at chromosome locations

13q34 and Xq22 respectively. Mutations in the α5(IV) gene have been associated with the X-linked kidney disorder Alport Syndrome and deletions involving both the α5(IV) and α6(IV) paired genes have been associated with Alport Syndrome accompanied by inherited smooth muscle tumours [5, 10–15].

References

1 Hostikka, S.L. and Tryggvason, K. (1987) Extensive structural differences between genes for the α1 and α2 chains of type IV collagen despite conservation of coding sequences. FEBS Lett. 224: 297–305.

2 Soininen, R. et al (1987) Complete primary structure of the α1-chain of human basement membrane (type IV) collagen. FEBS Lett. 225: 188–194.

3 Hostikka, S.L. and Tryggvason, K. (1988) The complete structure of the α2 chain of human type IV collagen and comparison with the α1(IV) chain. J. Biol. Chem. 263: 19488–19493.

4 Vandenberg, P. et al (1991) Characterization of a type IV collagen major cell binding site with affinity to the α1β1 and α2β1 integrins. J. Cell Biol. 113: 1475–1483.

5 Zhou, J. et al (1992) Complete amino acid sequence of the human α5(IV) collagen chain and identification of a single-base mutation in exon 23 converting glycine 521 in the collagenous domain to cysteine in an Alport Syndrome patient. J. Biol. Chem. 267: 12475–12481.

6 Furuto, D.K. and Miller, E.J. (1987) Isolation and characterization of collagens and procollagens. Methods Enzymol. 144: 41–61.

7 Morrison, K.E. et al (1991) Use of the polymerase chain reaction to clone and sequence a cDNA encoding the bovine α3 chain of type IV collagen. J. Biol. Chem. 266: 34–39.

8 Morrison, K.E. et al (1991) Sequence and localization of a partial cDNA encoding the human α3 chain of type IV collagen. Am. J. Hum. Genet. 49: 545–554.

9 Mariyama, M. et al (1992) The α4(IV) chain of basement membrane collagen. Isolation of cDNAs encoding bovine α4(IV) collagen and comparison with other type IV collagens. J. Biol. Chem. 267: 1253–1258.

10 Zhou, J. et al (1993) Deletion of the paired α5(IV) and α6(IV) collagen genes in inherited smooth muscle tumors. Science 261: 1167–1169.

11 Soininen, R. et al (1988) The structural genes for the α1 and α2 chains of type IV collagen are divergently encoded on opposite DNA strands and have an overlapping promoter region. J. Biol. Chem. 263: 17217–17220.

12 Soininen, R. et al (1989) Structural organisation of the gene for the α1(IV) chain of human type IV collagen. J. Biol. Chem. 264: 13565–13571.

13 Barker, D.F. et al (1990) Identification of mutations in the COL4A5 collagen gene in Alport syndrome. Science 248: 1224–1227.

14 Hostikka, S.L. et al (1990) Identification of a distinct type IV collagen α chain with a restricted kidney distribution and assignment of the gene to the locus of X-chromosome-linked Alport syndrome. Proc. Natl Acad. Sci. USA 87: 1606–1610.

15 Mariyama, M. et al (1992) Colocalisation of the genes for the α3(IV) and the α4(IV) chains of type IV collagen to chromosome 2 bands q35-q37. Genomics 13: 809–813.

Collagen type V

Type V collagen is a quantitatively minor fibrillar collagen and forms heterotypic fibrils with type I or types I and III collagens (and probably collagen types XII and XIV) in bone, tendon, cornea, skin, blood vessels and the more compliant tissues, liver, lung and placenta. It interacts preferentially over other fibrillar collagens with thrombospondin.

Molecular structure

Type V collagen is synthesized in both homotrimeric and heterotrimeric procollagen forms. The major form comprises two proα1(V) chains and one proα2(V) chain but in some tissues (notably uterus and placenta) a second form, comprising proα1(V), α2(V) and α3(V) chains, is present. A homotrimer of proα1(V) chains has been observed in cell culture of lung fibroblasts. In addition, the pro-α1(XI) chain can substitute for the proα1(V) chain forming cross-type heterotrimers. The procollagen forms are processed extracellularly by C-proteinase but only partial processing occurs at the NH_2-terminal end, resulting in part of the N-propeptide being retained in the extracellular matrix form. Type V collagen forms thin fibrils, and may control the diameter of type I collagen fibrils as a co-polymerized constituent [1–8].

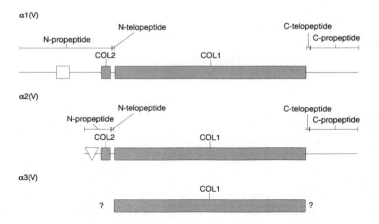

Isolation

Type V collagen can be prepared from foetal skin as partially processed forms by NaCl extraction [9], or as the shortened triple-helical form by pepsin digestion of placental tissues [10].

Accession number

P20908

Primary structure: α1(V) chain

Sequence conflicts:
641 E to G
650 P to L
663 R to E
668 E to Q
677 K to Q
684 L to P
692–699 PPGPPGVT to VTGEPGAP
727 G to Q
741 P to L
747 L to Q
753 P to A
759 D to N
776–777 GQ to QK
849–855 GGPNGDP to IGPPGPR
894 N to D

Ala	A	92	Cys	C	12	Asp D 105	Glu E 118
Phe	F	39	Gly	G	428	His H 17	Ile I 52
Lys	K	98	Leu	L	97	Met M 23	Asn N 34
Pro	P	334	Gln	Q	74	Arg R 72	Ser S 72
Thr	T	70	Val	V	54	Trp W 7	Tyr Y 40

Mol. wt (calc.) = 183 415   Residues = 1838

```
1     MDVHTRWKAR  SALRPGAPLL  PPLLLLLLWA  PPPSRAAQPA  DLLKVLDFHN
51    LPDGITKTTG  FCATRRSSKG  PDVAYRVTKD  AQLSAPTKQL  YPASAFPEDF
101   SILTTVKAKK  GSQAFLVSIY  NEQGIQQIGL  ELGRSPVFLY  EDHTGKPGPE
151   DYPLFRGINL  SDGKWHRIAL  SVHKKNVTLI  LDCKKKTTKF  LDRSDHPMID
201   INGIIVFGTR  ILDEEVFEGD  IQQLLFVSDH  RAAYDYCEHY  SPDCDTAVPD
251   TPQSQDPNPD  EYYTEGDGEG  ETYYYEYPYY  EDPEDLGKEP  TPSKKPVEAA
301   KETTEVPEEL  TPTPTEAAPM  PETSEGAGKE  EDVGIGDYDY  VPSEDYYTPS
351   PYDDLTYGEG  EENPDQPTDP  GAGAEIPTST  ADTSNSSNPA  PPPGEGADDL
401   EGEFTEETIR  NLDENYYDPY  YDPTSSPSEI  GPGMPANQDT  IYEGIGGPRG
451   EKGQKGEPAI  IEPGMLIEGP  PGPEGPAGLP  GPPGTMGPTG  QVGDPGERGP
501   PGRPGLPGAD  GLPGPPGTML  MLPFRFGGGG  DAGSKGPMVS  AQESQAQAIL
551   QQARLALRGP  AGPMGLTGRP  GPVGPPGSGG  LKGEPGDVGP  QGPRGVQGPP
601   GPAGKPGRRG  RAGSDGARGM  PGQTGPKGDR  GFDGLAGLPG  EKGHRGDPGP
651   SGPPGPPGDD  GERGDDGEVG  PRGLPGKPGP  RGLLGPKGPP  GPPGPPGVTG
701   MDGQPGPKGN  VGPQGEPGPP  GQQGNPGAQG  LPGPQGAIGP  PGEKGPLGKP
751   GLPGMPGADG  PPGHPGKEGP  PGEKGGQGPP  GPQGPIGYPG  PRGVKGADGI
801   RGLKGTKGEK  GEDGFPGFKG  DMGIKGDRGE  IGPPGPRGED  GPEGPKGRGG
851   PNGDPGPLGP  PGEKGKLGVP  GLPGYPGRQG  PKGSIGFPGF  PGANGEKGGR
901   GTPGKPGPRG  QRGPTGPRGE  RGPRGITGKP  GPKGNSGGDG  PAGPPGERGP
951   NGPQGPTGFP  GPKGPPGPPG  KDGLPGHPGQ  RGETGFQGKT  GPPGPPGVVG
1001  PQGPTGETGP  MGERGHPGPP  GPPGEQGLPG  LAGKEGTKGD  PGPAGLPGKD
1051  GPPGLRGFPG  DRGLPGPVGA  LGLKGNEGPP  GPPGPAGSPG  ERGPAGAAGP
1101  IGIPGRPGPQ  GPPGPAGEKG  APGEKGPQGP  AGRDGLQGPV  GLPGPAGPVG
1151  PPGEDGDKGE  IGEPGQKGSK  GDKGEQGPPG  PTGPQGPIGQ  PGPSGADGEP
1201  GPRGQQGLFG  QKGDEGPRGF  PGPPGPVGLQ  GLPGPPGEKG  ETGDVGQMGP
1251  PGPPGPRGPS  GAPGADGPQG  PPGGIGNPGA  VGEKGEPGEA  GEPGPSGRSG
```

47

```
1301    PPGPKGERGE    KGESGPSGAA    GPPGPKGPPG    DDGPKGSPGP    VGFPGDPGPP
1351    GEPGPAGQDG    PPGDKGDDGE    PGQTGSPGPT    GEPGPSGPPG    KRGPPGPAGP
1401    EGRQGEKGAK    GEAGLEGPPG    KTGPIGPQGA    PGKPGPDGLR    GIPGPVGEQG
1451    LPGSPGPDGP    PGPMGPPGLP    GLKGDSGPKG    EKGHPGLIGL    IGPPGEQGEK
1501    GDRGLPGPQG    SSGPKGEQGI    TGPSGPIGPP    GPPGLPGPPG    PKGAKGSSGP
1551    TGPRGEAGHP    GPPGPPGPPG    EVIQPLPIQA    SRTRRNIDAS    QLLDDGNGEN
1601    YVDYADGMEE    IFGSLNSLKL    EIEQMKRPLG    TQQNPARTCK    DLQLCHPDFP
1651    DGEYWVDPNQ    GCSRDSFKVY    CNFTAGGSTC    VFPDKKSEGA    RITSWPKENP
1701    GSWFSEFKRG    KLLSYVDAEG    NPVGVVQMTF    LRLLSASAHQ    NVTYHCYQSV
1751    AWQDAATGSY    DKALRFLGSN    DEEMSYDNNP    YIRALVDGCA    TKKGYQKTVL
1801    EIDTPKVEQV    PIVDIMFNDF    GEASQKFGFE    VGPACFMG
```

Structural and functional sites

Signal peptide: 1–37

N-Propeptide: 38–558

PARP repeat: 38–244

COL2 domain: 444–538 (interrupted triple helix of 5, 17 and 4 triplets similar to COL2 domains of proα2(V) and proα1(XI)

Helical domain: 559–1572

C-Telopeptide: 1573–1605

C-Propeptide: 1606–1838

Lysine/hydroxylysine cross-linking sites: 642, 1482

Potential N-linked glycosylation sites: 156, 1259, 1397

C-Proteinase cleavage site: 1605–1606

Accession number

P05997

Primary structure: α2(V) chain

Sequence conflicts: 1418 K to T
1438 F to S
1460–1496 E   to   Q
1496 V to A

Ala A 80	Cys C 17	Asp D 67	Glu E 75
Phe F 20	Gly G 403	His H 19	Ile I 37
Lys K 63	Leu L 66	Met M 24	Asn N 38
Pro P 269	Gln Q 59	Arg R 72	Ser S 62
Thr T 51	Val V 56	Trp W 6	Tyr Y 12

Mol. wt (calc.) = 144 560   Residues = 1496

```
1      MMANWAEARP    LLLILIVLLGQ    FVSIKAQEED    EDEGYGEEIA    CTQNGQMYLN
51     RDIWKPAPCQ    ICVCDNGAIL     CDKIECQDVL    DCADPVTPPG    ECCPVCSQTP
101    GGGNTNFGRG    RKGQKGEPGL     VPVVTGIRGR    PGPAGPPGSQ    GPRGERGPKG
151    RPGPRGPQGI    DGEPGVPGQP     GAPGPPGHPS    HPGPDGLSRP    FSAQMAGLDE
201    KSGLGSQVGL    MPGSVGPVGP     RGPQGLQGQQ    GGAGPTGPPG    EPGDPGPMGP
251    IGSRGPEGPP    GKPGEDGEPG     RNGNPGEVGF    AGSPGARGFP    GAPGLPGLKG
301    HRGHKGLEGP    KGEVGAPGSK     GEAGPTGPMG    AMGPLGPRGM    PGERGRLGPQ
351    GAPGQRGAHG    MPGKPGPMGP     LGIPGSSGFP    GNPGMKGEAG    PTGARGPEGP
401    QGQRGETGPP    GPVGSPGLPG     AIGTDGTPGP    KGPTGSPGTS    GPPGSAGPPG
```

451	SPGPQGSTGP	QGNSGLPGDP	GFKGEAGPKG	EPGPHGIQGP	IGPPGEEGKR
501	GPRGDPGTLG	PPGPVGERGA	PGNRGFPGSD	GLPGPKGAQG	ERGPVGSSGP
551	KGSQGDPGRP	GEPGLPGARG	LTGNPGVQGP	EGKLGPLGAP	GEDGRPGPPG
601	SIGIKGQPGT	MGLPGPKGSN	GDPGKPGEAG	NPGVPGQRGA	PGKDGKVGPY
651	GPPGPPGLRG	ERGEQGPPGP	TGFQGHPGPP	GPPGEGGKPG	DQGVPGGPGA
701	VGPLGPRGER	GNPGERGEPG	ITGLPGEKGM	AGGHGPDGPK	GSPGPSGTPG
751	DTGPPGLQGM	PGERGIAGTP	GPKGDRGGIG	EKGAEGTAGN	DGAGGLPGPL
801	GPPGPAGLLG	EKGEPGPRGL	VGPPGSRGNP	GSRGENGPTG	AVGFAGPQGS
851	DGQPGVKGEP	GEPGQKGDAG	SPGPQGLAGS	PGPHGPNGVP	GLKGGRGTQG
901	PPGATGFPGS	AGRVGPPGPA	GAPGPAGPLG	EPGKEGPPGP	RGDPGSHGRV
951	GVRGPAGPPG	GPGDKGDPGE	DGQPGPDGPP	GPAGTTGQRG	IVGMPGQRGE
1001	RGMPGLPGPA	GTPGKVGPTG	ATGDKGPPGP	VGPPGSNGPV	GEPGPEGPAG
1051	NDGTPGRDGA	VGERGDRGDP	GPAGLPGSQG	APGTPGPVGA	PGDAGQRGDP
1101	GSRGPIGHLG	RAGKRGLPGP	QGPRGDKGDH	GDRGDRGQKG	HRGFTGLQGL
1151	PGPPGPNGEQ	GSAGIPGPFG	PRGPPGPVGP	SGKEGNPGPL	GPLGPPGVRG
1201	SVGEAGPEGP	PGEPGPPGPP	GPPGHLTAAL	GDIMGHYDES	MPDPLPEFTE
1251	DQAAPDDKNK	TDPGVHATLK	SLSSQIETMR	SPDGSKKHPA	RTCDDLKLCH
1301	SAKQSGEYWI	DPNQGSVEDA	IKVYCNMETG	ETCISANPSS	VPRKTWWASK
1351	SPDNKPVWYG	LDMNRGSQFA	YGDHQSPNTA	ITQMTFLRLL	SKEASQNITY
1401	ICKNSVGYMD	DQAKNLKKAV	VLKGANDLDI	KAEGNIRFRY	IVLQDTCSKR
1451	NGNVGKTVFE	YRTQNVARLP	IIDLAPVDVG	GTDQEFGVEI	GPVCFV

Structural and functional sites

Signal peptide: 1–26

N-Propeptide: 27–212

von Willebrand factor C repeat: 36–104

COL2 domain: 110–188 (interrupted triple-helix of 4, 18 and 2 triplets similar to the COL2 domain in proα1(V), and proα1(XI))

Helical domain: 213–1223

C-Telopeptide: 1224–1226

C-Propeptide: 1227–1496

Lysine/hydroxylysine cross-linking sites: 201, 299, 1127

Potential N-linked glycosylation sites: 1259, 1397

C-Proteinase cleavage site: 1226–1227

Primary structure: α3(V) chain

Only the triple-helical domain has been sequenced (Accession: P25940) [11].

Gene structure

The proα1(V) and proα2(V) collagen chains are encoded by single genes found on human chromosomes 9 (locus q34.2–34.3) and 2 (locus q24.3–31), respectively. The proα2(V) gene is syntenic with the proα1(III) gene [5,12].

References

1 Fessler, J.H. and Fessler, L.I. (1987) Type V collagen. In: Structure and Function of Collagen Types, Mayne, R. and Burgeson, R.E., eds, Academic Press, London, pp. 81–103.

2 Weil, D. et al (1987) The pro-α2(V) collagen gene is evolutionarily related to the major fibril-forming collagens. Nucleic Acids Res. 15: 181–198.
3 Niyibizi,C. and Eyre,D.R. (1989) Identification of the cartilage α1(XI) chain in type V collagen from bovine bone. FEBS Lett. 242: 314–318.
4 Woodbury, D. et al (1989) Amino-terminal propeptide of human pro-α2(V) collagen conforms to the structural criteria of a fibrillar procollagen molecule. J. Biol. Chem. 264: 2735–2738.
5 Myers, J.C. and Dion, A.S. (1990) Types III and V procollagens: homology in genetic organization and diversity in structure. In: Extracellular Matrix Genes, Sandell, L.J. and Boyd, C.D., eds, Academic Press, New York, pp. 57–78.
6 Greenspan, D.S. et al (1991) The pro-α1(V) collagen chain. Complete primary structure, distribution of expression, and comparison with the pro-α1(XI) collagen chain. J. Biol. Chem. 266: 24727–24733.
7 Takahara, K. et al (1991) Complete primary structure of human collagen α1(V) chain. J. Biol. Chem. 266: 13124–13129.
8 Kleman, J.-P. et al (1992) The human rhabdomyosarcoma cell line A204 lays down a highly insoluble matrix composed mainly of α1 type-XI and α2 type-V collagen chains. Eur. J. Biochem. 210: 329–335.
9 Elstow, S.F. and Weiss, J.B. (1983) Extraction, isolation and characterization of neutral salt soluble type V collagen from foetal calf skin. Collagen Rel. Res. 3: 181–193.
10 Abedin, M.Z. et al (1982) Isolation and native characterization of cysteine-rich collagens from bovine placental tissues and uterus and their relationship to types IV and V collagens. Biosci. Rep. 2: 493–502.
11 Mann, K. (1992) Isolation of the α3-chain of human type V collagen and characterisation by partial sequencing. Hoppe-Seyler Z. Biol. Chem. 373: 69–75
12 Greenspan, D.S. et al (1992) Human collagen gene COL5A1 maps to the q34.2–q34.3 region of chromosome–9, near the locus for nail-patella syndrome. Genomics 12: 836–837.

Collagen type VI

Type VI collagen is essentially a glycoprotein with a short collagenous central domain and is present in most (if not all) connective tissues. It assembles into a unique supramolecular structure of 5 nm-diameter microfibrils with a periodicity of approximately 100 nm.

Molecular structure

Type VI collagen is synthesized largely as a heterotrimer comprising three genetically distinct α1(VI), α2(VI) and α3(VI) chains, although there is evidence for the existence of alternative but less stable assemblies comprised of either α3(VI) or α1(VI)/α2(VI) chains. Each chain consists of a short (105 nm) triple helix (approximately one-third the length of types I–III collagens) at each end of which is a large globular domain. The α1(VI) and α2(VI) chains are similar in size, but the α3(VI) chain is much larger due to its prominent NH2-terminal domain. Monomers are assembled intracellularly first into anti-parallel dimers and then into tetramers which are then secreted and assembled into microfibrils by end-to-end aggregation. The COOH-terminal domain of the α3(VI) chain is either unprocessed or only slightly processed. Type VI collagen microfibrils are stabilized by intra- and intermolecular disulphide bonds, but are not cross-linked by lysine/hydroxylysine-derived bonds.

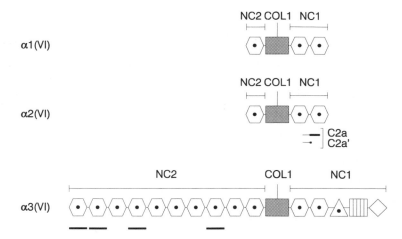

The non-collagenous domains NC1 and NC2 of the α1(VI) and α2(VI) chains are comprised of two and one repeats, respectively, that are homologous to the A-type repeats of von Willebrand factor. The NC1 domain of the α3(VI) chain also contains two A repeats as well as three other repeats; a lysine/proline-rich repeat homologous to several salivary proteins and containing threonine repeats which may be O-glycosylated, a fibronectin type III repeat and a repeat that is highly homologous to Kunitz-type serine proteinase inhibitors. The NC2 domain of the α3(VI) chain is comprised of at least ten von Willebrand factor A repeats. Multiple alternative splicing occurs at the NH2-terminal NC2 domain of the α3(VI) chain giving rise to at least four mutually exclusive isoforms. Three additional variants at the COOH-terminal end of the α2(VI) chain also arise by alternative splicing involving mutually exclusive use of the two most COOH-terminal exons together with the selective use of an internal acceptor splice site in the penultimate exon [1-8].

Isolation

Type VI collagen can be extracted completely and in an intact form from most tissues by selective glycosidase and/or guanidine–HCl treatment or bacterial collagenase digestion [9,10].

Accession number

S05377; A31952

Primary structure: α1(VI) chain

Sequence conflicts: 35 P to R
321–322 EK to QE

Ala	A	78	Cys	C	20	Asp	D	71	Glu	E	69
Phe	F	33	Gly	G	155	His	H	14	Ile	I	42
Lys	K	54	Leu	L	71	Met	M	11	Asn	N	28
Pro	P	92	Gln	Q	41	Arg	R	61	Ser	S	50
Thr	T	41	Val	V	65	Trp	W	4	Tyr	Y	28

Mol. wt (calc.) = 108 522   Residues = 1028

```
   1    MRAARALLPL  LLQACWTAAQ  DEPETPRAVA  FQDCPVDLFF  VLDTSESVAL
  51    RLKPYGALVD  KVKSFTKRFI  DNLRDRYYRC  DRNLVWNAGA  LHYSDEVEII
 101    QGLTRMPGGR  DALKSSVDAV  KYFGKGTYTD  CAIKKGLEQL  LVGGSHLKEN
 151    KYLIVVTDGH  PLEGYKEPCG  GLEDAVNEAK  HLGVKVFSVA  ITPDHLEPRL
 201    SIIATDHTYR  RNFTAADWGQ  SRDAEEAISQ  TIDTIVDMIK  NNVEQVCCSF
 251    ECQPARGPPG  LRGDPGFEGE  RGKPGLPGEK  GEAGDPGRPG  DLGPVGYQGM
 301    KGEKGSRGEK  GSRGPKGYKG  EKGKRGIDGV  DGVKGEMGYP  GLPGCKGSPG
 351    FDGIQGPPGP  KGDPGAFGLK  GEKGEPGADG  EAGRPGARGP  SGDEGPAGEP
 401    GPPGEKGEAG  DEGNPGPDGA  PGERGGPGER  GPRGTPGTRG  PRGDPGEAGP
 451    QGDQGREGPV  GVPGDPGEAG  PIGPKGYRGD  EGPPGSEGAR  GAPGPAGPPG
 501    DPGLMGERGE  DGPAGNGTEG  FPGFPGYPGN  RGAPGINGTK  GYPGLKGDEG
 551    EAGDPGDDNN  DIAPRGVKGA  KGYRGPEGPQ  GPPGHQGPPG  PDECEILDII
 601    MKMCSCCECK  CGPIDLLFVL  DSSESIGLQN  FEIAKDFVVK  VIDRLSRDEL
 651    VKFEPGQSYA  GVVQYSHSQM  QEHVSLRSPS  IRNVQELKEA  IKSLQWMAGG
 701    TFTGEALQYT  RDQLLPPSPN  NRIALVITDG  RSDTQRDTTP  LNVLCSPGIQ
 751    VVSVGIKDVF  DFIPGSDQLN  VISCQGLAPS  QGRPGLSLVK  ENYAELLEDA
 801    FLKNVTAQIC  IDKKCPDYTC  PITFSSPADI  TILLEPPPDV  GSHNFDTTKR
 851    FAKRLAERFL  TAGRTDPAHD  VRVAVVQYSG  TGQQRPERAS  LQFLQNYTAL
 901    ASAVDAMDFI  NDATDVNDAL  GYVTRFYREA  SSGAAKKRLL  LFSDGNSQGA
 951    TPAAIEKAVQ  EAQRAGIEIF  VVVVGRQVNE  PHIRVLVTGK  TAEYDVAYGE
1001    SHLFRVPSYQ  ALLRGVFHQT  VSRKVALG
```

Structural and functional sites

Signal peptide: 1–19 (probable)
NC2 domain: 20–256
Helical domain: 257–592
NC1 domain: 593–1028

von Willebrand factor A repeats: 30–216, 609–783, 801–1003

Hydroxylysine glycosylation sites: All residues in the Y position of Gly–X–Y triplets

Potential N-linked glycosylation sites: 212, 516 (determined), 537, 804, 896

Imperfections in Gly–X–Y triplets: 515–516, 559–565 (required for supercoiling of dimers)

Cysteine involved in dimer formation: 345

Accession number
B31952; S05378; S09646

Primary structure: α2(VI) chain

Sequence conflicts: 606 C to L
619 I to F
628 T to L
966 E to Q

Major variant:

Ala	A	60	Cys	C	21	Asp	D	70	Glu	E	66
Phe	F	39	Gly	G	154	His	H	21	Ile	I	42
Lys	K	51	Leu	L	66	Met	M	13	Asn	N	33
Pro	P	89	Gln	Q	44	Arg	R	67	Ser	S	54
Thr	T	45	Val	V	60	Trp	W	5	Tyr	Y	18

Mol. wt (calc.) = 108 354   Residues = 1018

C2a variant:

Ala	A	52	Cys	C	20	Asp	D	63	Glu	E	63
Phe	F	32	Gly	G	145	His	H	13	Ile	I	40
Lys	K	51	Leu	L	55	Met	M	12	Asn	N	26
Pro	P	91	Gln	Q	40	Arg	R	55	Ser	S	42
Thr	T	45	Val	V	48	Trp	W	6	Tyr	Y	18

Mol. wt (calc.) = 97 219   Residues = 917

C2a' variant:

Ala	A	42	Cys	C	20	Asp	D	57	Glu	E	55
Phe	F	27	Gly	G	142	His	H	12	Ile	I	36
Lys	K	49	Leu	L	46	Met	M	11	Asn	N	25
Pro	P	82	Gln	Q	35	Arg	R	50	Ser	S	41
Thr	T	35	Val	V	42	Trp	W	4	Tyr	Y	16

Mol. wt (calc.) = 87 092   Residues = 827

```
  1    MLQGTCSVLL   LWGILGAIQA   QQQEVISPDT   TERNNNCPEK   TDCPIHVYFV
 51    LDTSESVTMQ   SPTDILLFHM   KQFVPQFISQ   LQNEFYLDQV   ALSWRYGGLH
101    FSDQVEVFSP   PGSDRASFIK   NLQGISSFRR   GTFTDCALAN   MTEQIRQDRS
151    KGTVHFAVVI   TDGHVTGSPC   GIKLQAERAR   EEGIRLFAVA   PNQNLKEQGL
201    RDIASTPHEL   YRNDYATMLP   DSTEINQDTI   NRIIKVMKHE   AYGECYKVSC
251    LEIPGPSGPK   GYRGQKGAKG   NMGEPGEPGQ   KGRQGDPGIE   GPIGFPGPKG
301    VPGFKGEKGE   FGADGRKGAP   GLAGKNGTDG   QKGKLGRIGP   PGCKGDPGNR
351    GPDGYPGEAG   SPGERGDQGG   KGDPGRPGRR   GPPGEIGAKG   SKGYQGNNGA
```

```
401   PGSPGVKGAK   GGPGPRGPKG   EPGRRGDPGT   KGSPGSDGPK   GEKGDPGPEG
451   PRGLAGEVGN   KGAKGDRGLP   GPRGPQGALG   EPGKQGSRGD   PGDAGPRGDS
501   GQPGPSGDPG   RPGFSYPGPR   GAPGEKGEPG   PRGPEGGRGD   FGLKGEPGRK
551   GEKGEPADPG   PPGEPGPRGP   RGVPGPEGEP   GPPGDPGLTE   CDVMTYVRET
601   CGCCDCEKRC   GALDVVFVID   SSESIGYTNF   TLEKNFVINV   VNRLGAIAKD
651   PKSETGTRVG   VVQYSHEGTF   EAIQLDDEHI   DSLSSFKEAV   KNLEWIAGGT
701   WTPSALKFAY   DRLIKESRRQ   KTRVFAVVIT   DGRHDPRDDD   LNLRALCDRD
751   VTVTAIGIGD   MFHEKHESEN   LYSIACDKPQ   QVRNMTLFSD   LVAEKFIDDM
801   EDVLCPDPQI   VCPDLPCQT
```

Major variant continues:
```
820                        E   LSVAQCTQRP   VDIVFLLDGS   ERLGEQNFHK
850   ARRFVEQVAR   RLTLARRDDD   PLNARVALLQ   FGGPGEQQVA   FPLSHNLTAI
901   HEALETTQYL   NSFSHVGAGV   VHAINAIVRS   PRGGARRHAE   LSFVFLTDGV
951   TGNDSLHESA   HSMRNENVVP   TVLALGSDVD   MDVLTTLSLG   DRAAVFHEKD
1001  YDSLAQPGFF   DRFIRWIC
```

C2a variant continues:
```
820                        D   APWPGGEPPV   TFLRTEEGPD   ATFPRTIPLI
850   QQLLNATELT   QDPAAYSQLV   AVLVYTAERA   KFATGVERQD   WMELFIDTFK
901   LVHRDIVGDP   ETALALC
```

C2a' variant continues:
```
820                        G   LDGAVLC
```

Structural and functional sites

Signal peptide: 1–20
NC2 domain: 21–254
Helical domain: 255–589
NC1 domain: 590–1018 (main variant); 590–917 (C2a variant); 590–827 (C2a' variant)
von Willebrand factor A repeats: 36–218, 606–783, 818–995
Hydroxylysine glycosylation sites: All residues in the Y position of Gly–X–Y triplets
Potential N-linked glycosylation sites: 140 (determined), 326, 629 (determined), 784, 855 (C2a variant only), 896, 953
Imperfections in Gly–X–Y triplets: 516–517, 557–559 (required for supercoiling of dimers)
Cysteine involved in dimer formation: 343

Accession number
S13679

Primary structure: α3(VI) chain

Ala	A	234	Cys	C	30	Asp	D	180	Glu	E	183

Ala A 234   Cys C 30   Asp D 180   Glu E 183
Phe F 150   Gly G 303   His H 43   Ile I 154
Lys K 158   Leu L 283   Met M 34   Asn N 130
Pro P 211   Gln Q 146   Arg R 189   Ser S 215
Thr T 169   Val V 295   Trp W 6   Tyr Y 62
Mol. wt (calc.) = 342 960   Residues = 3175

1	MRKHRHLPLV	AVFCLFLSGF	PTTHAVKNGA	AADIIFLVDS	SWTIGEEHFQ
51	LVREFLYDVV	KSLAVGENDF	HFALVQFNGN	PHTEFLLNTY	RTKQEVLSHI
101	SNMSYIGGTN	QTGKGLEYIM	AKHLTKAAGS	LAGDGVPQVI	VVLTDGHSKD
151	GLALPSAELK	SADVNVFAIG	VEDADEGALK	EIASEPLNMH	MFNLENFTSL
201	HDIVGNLVSC	VHSSVSPERA	GDTETLKDIT	QQQQAAQDSA	DIIFLIDGSN
251	NTGSVNFAVI	LDFLVNLLEK	LPIGTQQIRV	GVVQFSDEPR	TMFSLDTYST
301	KAQVLGAVKA	LGFAGGELAN	IGLALDFVVE	NHFTRAGGSR	VEEGVPQVLV
351	LISAGPSSDE	IRYGVVALKQ	ASVFSFGLGA	QAASRAELQH	IATDDNLVFT
401	VPEFRSFGDL	QEKLLPYIVG	VAQRHIVLKP	PTIVTQVIEV	NKRDIVFLVD
451	GSSALGLANF	NAIRDFIAKV	IQRLEIGQDL	IQVAVAQYAD	TVRPEFYFNT
501	HPTKREVITA	VRKMKPLDGS	ALYTGSALDF	VRNNLFTSSA	GYRAAEGIPK
551	LLVLITGGKS	LDEISQPAQE	LKRSSIMAFA	IGNKGADQAE	LEEIAFDSSL
601	VFIPAEFRAA	PLQGMLPGLL	APLRTLSGTP	EVHSNKRDII	FLLDGSANVG
651	KTNFPYVRDF	VMNLVNSLDI	GNDNIRVGLV	QFSDTPVTEF	SLNTYQTKSD
701	ILGHLRQLQL	QGGSGLNTGS	ALSYVYANHF	TEAGGSRIRE	HVPQLLLLLT
751	AGQSEDSYLQ	AANALTRAGI	LTFCVGASQA	NKAELEQIAF	NPSLVYLMDD
801	FSSLPALPQQ	LIQPLTTYVS	GGVEEVPLAQ	PESKRDILFL	FDGSANLVGQ
851	FPVVRDFLYK	IIDELNVKPE	GTRIAVAQYS	DDVKVESRFD	EHQSKPEILN
901	LVKRMKIKTG	KALNLGYALD	YAQRYIFVKS	AGSRIEDGVL	QFLVLLVAGR
951	SSDRVDGPAS	NLKQSGVVPF	IFQAKNADPA	ELEQIVLSPA	FILAAESLPK
1001	IGDLHPQIVN	LLKSVHNGAP	APVSGEKDVV	FLLDGSEGVR	SGFPLLKEFV
1051	QRVVESLDVG	QDRVRVAVVQ	YSDRTRPEFY	LNSYMNKQDV	VNAVRQLTLL
1101	GGPTPNTGAA	LEFVLRNILV	SSAGSRITEG	VPQLLIVLTA	DRSGDDVRNP
1151	SVVVKRGGAV	PIGIGIGNAD	ITEMQTISFI	PDFAVAIPTF	RQLGTVQQVI
1201	SERVTQLTRE	ELSRLQPVLQ	PLPSPGVGGK	RDVVFLIDGS	QSAGPEFQYV
1251	RTLIERLVDY	LDVGFDTTRV	AVIQFSDDPK	AEFLLNAHSS	KDEVQNAVQR
1301	LRPKGGRQIN	VGNALEYVSR	NIFKRPLGSR	IEEGVPQFLV	LLISSGKDDE
1351	VVVPAVELKQ	FGVAPFTIAR	NADQEELVKI	SLSPEYVFSV	STFRELPSLE
1401	QKLLTPITTL	TSEQIQKLLA	STRYPPPAVE	SDAADIVFLI	DSSEGVRPDG
1451	FAHIRDFVSR	IVRRLNIGPS	KVRVGVVQFS	NDVFPEFYLK	TYRSQAPVLD
1501	AIRRLRLRGG	SPLNTGKALE	FVARNLFVKS	AGSRIEDGVP	QHLVLVLGGK
1551	SQDDVSRFAQ	VIRSSGIVSL	GVGDRNIDRT	ELQTITNDPR	LVFTVREFRE
1601	LPNIEERIMN	SFGPSAATPA	PPGVDTPPPS	RPEKKKADIV	FLLDGSINFR
1651	RDSFQEVLRF	VSEIVDTVYE	DGDSIQVGLV	QYNSDPTDEF	FLKDFSTKRQ
1701	IIDAINKVVY	KGGRHANTKV	GLEHLRVNHF	VPEAGSRLDQ	RVPQIAFVIT
1751	GGKSVEDAQD	VSLALTQRGV	KVFAVGVRNI	DSEEVGKIAS	NSATAFRVGN
1801	VQELSELSEQ	VLETFDDAID	ETLCPGVTDA	AKACNLDVIL	GFDGSRDQNV
1851	FVAQKGFESK	VDAILNRISQ	MHRVSCSGGR	SPTVRVSVVA	NTPSGPVEAF
1901	DFDEYQPEML	EKFRNMRSQH	PYVLTEDTLK	VYLNKFRQSS	PDSVKVVIHF
1951	TDGADGDLAD	LHRASENLRQ	EGVRALILVG	LERVVNLERL	MHLEFGRGFM
2001	YDRPLRLNLL	DLDYELAEQL	DNIAEKACCG	VPCKCSGQRG	DRGPIGSIGP
2051	KGIPGEDGYR	GYPGDEGGPG	ERGPPGVNGT	QGFQGCPGQR	GVKGSRGFPG
2101	EKGEVGEIGL	DGLDGEDGDK	GLPGSSGEKG	NPGRRGDKGP	RGEKGERGDV
2151	GIRGDPGNPG	QDSQERGPKG	ETGDLGPMGV	PGRDGVPGGP	GETGKNGGFG
2201	RRGPPGAKGN	KGGPGQPGFE	GEQGTRGAQG	PAGPAGPPGL	IGEQGISGPR
2251	GSGGARGAPG	ERGRTGPLGR	KGEPGEPGPK	GGIGNPGPRG	ETGDDGRDGV
2301	GSEGRRGKKG	ERGFPGYPGP	KGNPGEPGLN	GTTGPKGIRG	RRGNSGPPGI
2351	VGQKGRPGYP	GPAGPRGNRG	DSIDQCALIQ	SIKDKCPCCY	GPLECPVFPT
2401	ELAFALDTSE	GVNQDTFGRM	RDVVLSIVNV	LTIAESNCPT	GARVAVVTYN
2451	NEVTTEIRFA	DSKRKSVLLD	KIKNLQVALT	SKQQSLETAM	SFVARNTFKR
2501	VRNGFLMRKV	AVFFSNTPTR	ASPQLREAVL	KLSDAGITPL	FLTRQEDRQL

```
2551  INALQINNTA  VGHALVLPAG  RDLTDFLENV  LTCHVCLDIC  NIDPSCGFGS
2601  WRPSFRDRRA  AGSDVDIDMA  FILDSAETTT  LFQFNEMKKY  IAYLVRQLDM
2651  SPDPKASQHF  ARVAVVQHAP  SESVDNASMP  PVKVEFSLTD  YGSKEKLVDF
2701  LSRGMTQLQG  TRALGSAIEY  TIENVFESAP  NPRDLKIVVL  MLTGEVPEQQ
2751  LEEAQRVILQ  AKCKGYFFVV  LGIGRKVNIK  EVYTFASEPN  DVFFKLVDKS
2801  TELNEEPLMR  FGRLLPSFVS  SENAFYLSPD  IRKQCDWFQG  DQPTKNLVKF
2851  GHKQVNVPNN  VTSSPTSNPV  TTTKPVTTTK  PVTTTTKPVT  TTTKPVTIIN
2901  QPSVKPAAAK  PAPAKPVAAK  PVATKTATVR  PPVAVKPATA  AKPVAAKPAA
2951  VRPPAAAAKP  VATKPEVPRP  QAAKPAATKP  ATTKPVVKML  REVQVFEITE
3001  NSAKLHWERP  EPPGPYFYDL  TVTSAHDQSL  VLKQNLTVTD  RVIGGLLAGQ
3051  TYHVAVVCYL  RSQVRATYHG  SFSTKKSQPP  PPQPARSASS  STINLMVSTE
3101  PLALTETDIC  KLPKDEGTCR  DFILKWYYDP  NTKSCARFWY  GGCGGNENKF
3151  GSQKECEKVC  APVLAKPGVI  SVMGT
```

Structural and functional sites

Signal peptide: 1–25
NC2 domain: 26–2036
Helical domain: 2037–2372
NC1 domain: 2373–3175
von Willebrand factor A repeats: 26–230, 239–425, 432–632, 636–827, 834–1020, 1026–1212, 1230–1419, 1433–1625, 1636–1823, 1835–2027, 2399–2582, 2616–2863
Lysine/proline-rich repeat: 2864–2985
Fibronectin type III repeat: 2986–3074
Kunitz-type serine proteinase inhibitor repeat: 3110–3160
Alternatively spliced repeats: 1433–1625 (N3), 636–827 (N7), 239–425 (N9), 26–230 (N10)
Hydroxylysine glycosylation sites: All residues in the Y position of Gly–X–Y triplets
Potential N-linked glycosylation sites: 102, 110, 196, 250, 791, 1149, 2078, 2330, 2557, 2676, 2860, 3035
Imperfections in Gly–X–Y triplets: 2163–2166, 2299–2300 (required for supercoiling of dimers)
Cysteine involved in tetramer formation: 2086

Gene structure

The human $\alpha1(VI)$ collagen gene is located on chromosome 21 (locus q22.3). The $\alpha2(VI)$ collagen gene has a similar location, is approximately 35 kb in size, and contains 30 exons. The exon structure encoding the triple-helical domain of both the $\alpha1(VI)$ and $\alpha2(VI)$ chains is almost identical and based on multiples of 9 bp (27, 36, 45, 54, 63, and 90 bp) except for interrupted regions. The $\alpha3(VI)$ collagen gene is located on human chromosome 2 at locus q37. The exons encoding the NH$_2$-terminal domain alone span 26 kb [2,6,11].

References
[1] Timpl, R. and Engel, J. (1987) Type VI collagen. In: Structure and Function of Collagen Types, Mayne, R. and Burgeson, R.E., eds, Academic Press, London, pp. 105–153.

2 Weil, D. et al (1988) Cloning and chromosomal localization of human genes encoding the three chains of type VI collagen. Am. J. Hum. Genet. 42: 435–445.

3 Chu, M.-L. et al (1989) Sequence analysis of α1(VI) and α2(VI) chains of human type VI collagen reveals internal triplication of globular domains similar to the A domains of von Willebrand factor and two α2(VI) chain variants that differ in the carboxy terminus. EMBO J. 8: 1939–1946.

4 Chu, M.-L. et al (1990) Mosaic structure of globular domains in the human type VI collagen α3 chain: Similarity to von Willebrand factor, salivary proteins and aprotinin type protease inhibitors. EMBO J. 9: 385–393.

5 Saitta, B. et al (1990) Alternative splicing of the human α2(VI) collagen gene generates multiple mRNA transcripts which predict three protein variants with distinct carboxy termini. J. Biol. Chem. 265: 6473–6480.

6 Stokes, D.G. et al. (1991) Human α3(VI) collagen gene. Characterization of exons coding for the amino-terminal globular domain and alternative splicing in normal and tumor cells. J. Biol. Chem. 266: 8626–8633.

7 Saitta, B. et al (1992) Human α2(VI) collagen gene. Heterogeneity at the 5'-untranslated region generated by an alternative exon. J. Biol. Chem. 267: 6188–6196.

8 Zanussi, S. et al (1992) The human type VI collagen gene. mRNA and protein variants of the α3 chain generated by alternative splicing of an additional 5'-end exon. J. Biol. Chem. 267: 24082–24089.

9 Wu, J.-J. et al (1987) Type VI collagen of the intervertebral disc. Biochemical and electron-microscopic characterization of the native protein. Biochem. J. 248: 373–381.

10 Kielty, C.M. et al (1991) Isolation and ultrastructural analysis of microfibrillar structures from foetal bovine elastic tissues. Relative abundance and supramolecular architecture of type VI collagen assemblies and fibrillin. J. Cell Sci. 99: 797–807.

11 Saitta, B. et al (1991) The exon organization of the triple-helical coding regions of the human α1(VI) and α2(VI) collagen genes is highly similar. Genomics 11: 145–153.

Collagen type VII

Type VII collagen is one of the largest collagens and has a highly specific tissue distribution. It is the major component of anchoring fibrils that are elaborated by specialized epithelia (e.g. epidermis and intestinal submucosa) and which anchor the basement membrane to the underlying stromal tissue.

Molecular structure

Type VII collagen is a homotrimer of three α1(VII) chains. Each chain comprises a triple helix, approximately 424 nm in length, a large NH_2-terminal globular domain and a small COOH-terminal domain. The monomers assemble into antiparallel dimers in the tissue with a COOH- to COOH-terminal overlap of 60 nm, resulting in an overall triple helix of 785 nm. Limited processing at the COOH-terminal end appears to occur prior to dimerization. The dimers then associate laterally to produce "segment-long spacing (SLS)-like" structures that insert into the basement membrane via the large NH_2-terminal domains. The NH_2-terminal domain comprises repeats that are homologous to the A repeats of von Willebrand factor and there are nine consecutive fibronectin type III repeats. The small COOH-terminal domain is cysteine-rich and is homologous to the Kunitz proteinase inhibitor-type repeat found in the α3(VI) chain [1–8].

Isolation

Type VII collagen can be isolated in an intact form either from the culture medium of specialized epithelial cells or from guanidine–HCl extracts of amniotic membranes [2]. The triple-helical domain can be isolated in larger quantities by pepsin digestion [1].

Primary structure

Only the NC1 domain and parts of the NC2 and triple-helical domains have been sequenced (Accession: S16316, PH0844) [4–8]. N.B. While the COL and NC domains follow the usual convention and are numbered from C to N, it should be noted that, in the cited references, the domains are numbered from N to C.

Gene structure

The type VII collagen gene is located on the short arm of chromosome 3 at locus 3p21.3.

References

[1] Bentz, H. et al (1983) Isolation and partial characterization of a new human collagen with extended triple-helical structural domain. Proc. Natl Acad. Sci. USA 80: 3168–3172.

[2] Lunstrum, G.P. et al (1986) Large complex globular domains of type VII

procollagen contribute to the structure of anchoring fibrils. J. Biol. Chem. 261: 9042–9048.

[3] Burgeson, R.E. (1987) Type VII collagen. In: Structure and Function of Collagen Types, Mayne, R. and Burgeson, R.E., eds, Academic Press, London, pp. 145–172.

[4] Parente, M.G. et al. (1991) Human type VII collagen: cDNA cloning and chromosomal mapping of the gene. Proc. Natl Acad. Sci. USA 88: 6931–6935.

[5] Tanaka, T. et al (1992) Molecular cloning and characterization of type VII collagen cDNA. Biochem. Biophys. Res. Commun. 183: 958–963.

[6] Gammon, W.R. et al (1992) Noncollagenous (NC1) domain of collagen VII resembles multidomain adhesion proteins involved in tissue-specific organization of extracellular matrix. J. Invest. Dermatol. 99: 691–696.

[7] Christiano, A.M. et al (1992) The large non-collagenous domain (NC-1) of type VII collagen is amino-terminal and chimeric. Homology to cartilage matrix protein, the type III domains of fibronectin and the A domains of von Willebrand factor. Human Mol. Genet. 1: 475–481.

[8] Greenspan, D.S. (1993) The carboxyl-terminal half of type VII collagen, including the non-collagenous NC-2 domain and intron/exon organization of the corresponding region of the COL 7A1 gene. Human Mol. Genet. 2: 273–278.

Collagen type VIII

Type VIII collagen is a major component of Descemet's membrane, the specialized basement membrane elaborated by corneal endothelial cells, but it is also synthesized by vascular endothelial cells and epithelial and mesenchymal cells of other tissues. It assembles into the hexagonal lattice that is a characteristic feature of Descemet's membrane .

Molecular structure

Two α chains, α1(VIII) and α2(VIII), have been characterized and there is evidence to suggest that in Descemet's membrane the molecule is a heterotrimer comprising two α1(VIII) chains and one α2(VIII) chain. The COOH-terminal three-quarters of each NC1 domain is homologous to the corresponding region of the α1(X) chain. The lengths of the triple-helical and NC1 domains are similar to those of type X collagen, but the NC2 domain is two to three times larger [1–5].

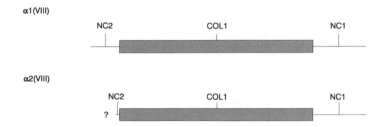

Isolation

Type VIII collagen can be isolated in an intact form from the medium of cultured endothelial cells [6] or as its triple-helical domain by pepsin digestion [7].

Accession number

S15435

Primary structure: α1(VIII) chain

Ala	A	31	Cys	C	2	Asp	D	11	Glu	E	26
Phe	F	18	Gly	G	191	His	H	12	Ile	I	33
Lys	K	47	Leu	L	53	Met	M	23	Asn	N	8
Pro	P	163	Gln	Q	38	Arg	R	14	Ser	S	13
Thr	T	9	Val	V	30	Trp	W	1	Tyr	Y	21

Mol. wt (calc.) = 73 358   Residues = 744

```
1    MAVLPGPLQL  LGVLLTISLS  SIRLIQAGAY  YGIKPLPPQI  PPQMPPQIPQ
51   YQPLGQQVPH  MPLAKDGLAM  GKEMPHLQYG  KEYPHLPQYM  KEIQPAPRMG
101  KEAVPKKGKE  IPLASLRGEQ  GPRGEPGPRG  PPGPPGLPGH  GIPGIKGKPG
151  PQGYPGVGKP  GMPGMPGKPG  AMGMPGAKGE  IGQKGEIGPM  GIPGPQGPPG
201  PHGLPGIGKP  GGPGLPGQPG  PKGDRGPKGL  PGPQGLRGPK  GDKGFGMPGA
251  PGVKGPPGMH  GLPGPVGLPG  VGKPGVTGFP  GPQGPLGKPG  APGEPGRQGP
301  IGVPGVQGPP  GIPGIGKPGQ  DGIPGQPGFP  GGKGEQGLPG  LPGAPGLPGI
```

351	GKPGFPGPKG	DRGMGGVPGA	LGPRGEKGPI	GSPGIGGSPG	EPGLPGIPGP
401	MGPPGAIGFP	GPKGEGGIVG	PQGPPGPKGE	PGLQGFPGKP	GFLGEVGPPG
451	MRGFPGPIGP	KGEHGQKGVP	GLPGVPGLLG	PKGEPGIPGD	QGLQGPPGIP
501	GIGGPSGPIG	PPGIPGPKGE	PGLPGPPGFP	GIGKPGVAGL	HGPPGKPGAL
551	GPQGQPGLPG	PPGPPGPPGP	PAVMPPTPPP	QGEYLPDMGL	GIDGVKPPHA
601	TGAKKGKNGG	PAYEMPAFTA	ELTAPFPPVG	GPVKFNKLLY	NGRQNYNPQT
651	GIFTCEVPGV	YYFAYHVHCK	GGNVWVALFK	NNEPVMYTYD	EYKKGFLDQA
701	SGSAVLLLRP	GDRVFLQMPS	EQAAGLYAGQ	YVHSSFSGYL	LYPM

Structural and functional sites

Signal peptide: 1–28
NC2 domain: 29–117
Helical domain: 118–571
NC1 domain: 572–744
Imperfections in Gly–X–Y triplets: 139–140, 156–157, 206–207, 244–245, 270–271, 314–315, 349–350, 531–532 (same relative locations as imperfections in α2(VIII) and α1(X))
Mammalian collagenase cleavage sites: 206–207, 531–532

Primary structure: α2(VIII) chain

Only the sequence of the triple-helical domain and the COOH-terminal domain has been determined (Accession: P25067) [8].

Gene structure

The α1(VIII) collagen gene has been localized to the long arm of human chromosome 3 at locus q12–13.1. The α2(VIII) collagen gene is located on the short arm of chromosome 1 at locus p32.3–34.3. The genes encoding both α chains have a condensed structure similar to that of type X collagen. The gene for the rabbit α1(VIII) chain comprises four exons, exons 1 and 2 coding for the 5′-untranslated region, exon 3 encoding most of the NC2 domain and exon 4 encoding the remainder of the NC2 domain, the entire triple-helical domain, the NC1 domain and the 3′-untranslated region [3,8].

References

[1] Mann, K. et al. (1990) The primary structure of a triple-helical domain of collagen type VIII from bovine Descemet's membrane. FEBS Lett. 273: 168–172.

[2] Sawada, H. et al (1990) Characterization of the collagen in the hexagonal lattice of Descemet's membrane: Its relation to type VIII collagen. J. Cell Biol. 110: 219–227.

[3] Muragaki, Y. et al (1991) The complete primary structure of the human α1(VIII) chain and assignment of its gene (COL8A1) to chromosome 3. Eur. J. Biochem. 197: 615–622.

[4] Yamaguchi, N. et al (1991) The α1(VIII) collagen gene is homologous to the α1(X) collagen gene and contains a large exon encoding the entire triple helical and carboxyl-terminal non-triple helical domains of the α1(VIII) polypeptide. J. Biol. Chem. 266: 4508–4513.

[5] Muragaki, Y. et al (1992) α(VIII) collagen gene transcripts encode a short chain

collagen polypeptide and are expressed by various epithelial, endothelial and mesenchymal cells in newborn mouse tissues. Eur. J. Biochem. 207: 895–902.

6 Benya, P.D. and Padilla, S.R. (1986) Isolation and characterization of type VIII collagen synthesised by cultured rabbit corneal endothelial cells. A conventional structure replaces the interrupted-helix model. J. Biol. Chem. 261: 4160–4169.

7 Kapoor, R. et al (1986) Type VIII collagen from bovine Descemet's membrane: Structural characterization of a triple-helical domain. Biochemistry, 25: 3930–3937.

8 Muragaki, Y. et al (1991) The α2(VIII) collagen gene. A novel member of the short chain collagen family located on the human chromosome 1. J. Biol. Chem. 266: 7721–7727.

Collagen type IX

Type IX collagen is the prototype of a subfamily of collagens called FACIT collagens that include types XII, XIV and XVI. Type IX collagen associates specifically with the surface of, and participates in the formation of, type II collagen fibrils. It is found in cartilage, intervertebral disc and vitreous humour. As well as linking type II collagen molecules it may also serve to bridge collagen fibrils with other matrix macromolecules.

Molecular structure

Type IX collagen is synthesized as a disulphide-bonded heterotrimer comprising three distinct $\alpha1(IX)$, $\alpha2(IX)$ and $\alpha3(IX)$ chains. The molecule is not processed prior to its deposition in the extracellular matrix. The molecule comprises three collagenous domains (COL1–3) and four non-collagenous domains (NC1–4). The $\alpha2(IX)$ chain can have a chondroitin/dermatan sulphate glycosaminoglycan covalently attached at the NC3 domain; there are, therefore, both proteoglycan and non-proteoglycan forms of type IX collagen. The length of the glycosaminoglycan is both species- and tissue-dependent. Another form of type IX collagen lacks the large NC4 domain at the NH_2-terminus of the $\alpha1(IX)$ chain due to the use of an alternative promoter in the $\alpha1(IX)$ gene. The expression of this variant is tissue-specific and developmentally regulated.

The COL1 domain of type IX collagen is homologous to the COL1 domains of type XII, XIV and XVI collagens. The NC4 domain contains a PARP repeat and shows homology to the NC3 domain of types XII and XIV collagen. The NH_2-terminal region of the COL2 domain of all three chains is cross-linked to the N-telopeptides of type II collagen; the COL2 domain of $\alpha3(IX)$ is cross-linked to the C-telopeptide of type II collagen. The two cross-link sites on the $\alpha3(IX)$ chain span precisely the gap zone of type II collagen fibrils and as a consequence the type IX collagen molecules are anti-parallel to the type II collagen molecules in the fibrils. The $\alpha3(IX)$ NC1 domain also cross-links with the COL2 domain of $\alpha1(IX)$ and $\alpha3(IX)$ chains. The NC4 domain of the $\alpha1(IX)$ chain may interact with the glycosaminoglycan chains of proteoglycans [1–11].

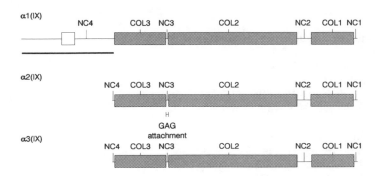

Isolation

The intact type IX molecule has been isolated from a rat chondrosarcoma cell line by NaCl extraction [12] or from foetal cartilage by guanidine–HCl extraction [13].

Large quantities of the cleaved COL1–3 domains are prepared by pepsin digestion of cartilage [14].

Accession number
P20849

Primary structure: α1(IX) chain

Ala	A	49	Cys	C	11	Asp	D	38	Glu	E	49
Phe	F	21	Gly	G	216	His	H	9	Ile	I	35
Lys	K	43	Leu	L	65	Met	M	12	Asn	N	19
Pro	P	149	Gln	Q	38	Arg	R	53	Ser	S	45
Thr	T	29	Val	V	37	Trp	W	7	Tyr	Y	6

Mol. wt (calc.) = 92 867   Residues = 931

```
1    MKTCWKIPVF  FFVCSFLEPW  ASAAVKRRPR  FPVNSNSNGG  NELCPKIRIG
51   QDDLPGFDLI  SQFQVDKAAS  RRAIQRVVGS  ATLQVAYKLG  NNVDFRIPTR
101  NLYPSGLPEE  YSFLTTFRMT  GSTLKKNWNI  WQIQDSSGKE  QVGIKINGQT
151  QSVVFSYKGL  DGSLQTAAFS  NLSSLFDSQW  HKIMIGVERS  SATLFVDCNR
201  IESLPIKPRG  PIDIDGFAVL  GKLADNPQVS  VPFELQWMLI  HCDPLRPRRE
251  TCHELPARIT  PSQTTDERGP  PGEQGPPGAS  GPPGVPGIDG  IDGDRGPKGP
301  PGPPGPAGEP  GKPGAPGKPG  TPGADGLTGP  DGSPGSIGSK  GQKGEPGVPG
351  SRGFPGRGIP  GPPGPPGTAG  LPGELGRVGP  VGDPGRRGPP  GPPGPPGPRG
401  TIGFHDGDPL  CPNACPPGRS  GYPGLPGMRG  HKGAKGEIGE  PGRQGHKGEE
451  GDQGELGEVG  AQGPPGAQGL  RGITGLVGDK  GEKGARGLDG  EPGPQGLPGA
501  PGDQGQRGPP  GEAGPKGDRG  AEGARGIPGL  PGPKGDTGLP  GVDGRDGIPG
551  MPGTKGEPGK  PGPPGDAGLQ  GLPGVPGIPG  AKGVAGEKGS  TGAPGKPGQM
601  GNSGKPGQQG  PPGEVGPRGP  QGLPGSRGEL  GPVGSPGLPG  KLGSLGSPGL
651  PGLPGPPGLP  GMKGDRGVVG  EPGPKGEQGA  SGEEGEAGER  GELGDIGLPG
701  PKGSAGNPGE  PGLRGPEGSR  GLPGVEGPRG  PPGPRGVQGE  QGATGLPGVQ
751  GPPGRAPTDQ  HIKQVCMRVI  QEHFAEMAAS  LKRPDSGATG  LPGRPGPPGP
801  PGPPGENGFP  GQMGIRGLPG  IKGPPGALGL  RGPKGDLGEK  GERGPPGRGP
851  NGLPGAIGLP  GDPGPASYGK  NGRDGERGPP  GLAGIPGVPG  PPGPPGLPGF
901  CEPASCTMQL  VSEHLTKGLT  LERLTAAWLS  A
```

Structural and functional sites
Signal peptide: 1–23
NC4 domain: 24–268
PARP repeat 28–267
COL3 domain: 269–405
NC3 domain: 406–417
COL2 domain: 418–756
Lysine/hydroxylysine cross-linking site: 432
NC2 domain: 757–786
COL1 domain: 787–901
NC1 domain: 902–931
Potential N-linked glycosylation site: 171
Interchain disulphide bond residues: 411, 415, 901, 906
Imperfections in Gly–X–Y triplets: 356–360, 847–851, 864–868
Stromelysin cleavage site: 780–781

Primary structure: α1(IX) chain (alternative short form)

```
1    MAWTARDRGA   LGLLLLGLCL   CAAQR
```

Structural and functional sites
Signal peptide: 1–23
NC4 domain: 24–25
COL3–NC1: 26–688 (identical in sequence to the long form)

Accession number
P12108

Primary structure: α2(IX) chain (chicken)

Ala	A	43	Cys	C	10	Asp	D 36	Glu	E 36
Phe	F	18	Gly	G	209	His	H 13	Ile	I 40
Lys	K	55	Leu	L	54	Met	M 15	Asn	N 12
Pro	P	134	Gln	Q	50	Arg	R 42	Ser	S 35
Thr	T	29	Val	V	43	Trp	W 5	Tyr	Y 9

Mol. wt (calc.) = 64 995   Residues = 677

```
1    MAHRSPALCL   LLLLHAACLCL   AQLRGPPGEP   GPRGPPGPPG   VPGADGIDGD
51   KGSPGAPGSP   GAKGEPGAPG    PDGPPGKPGL   DGLTGAKGSR   GPWGGQGLKG
101  QPGLPGPPGL   PGPSLPGPPG    LPGQVGLPGE   IGVPGPKGDP   GPDGPRGPPG
151  PPGKPGPPGH   IQGVEGSADF    LCPTNCPPGP   KGPQGLQGLK   GHRGRPGALG
201  EPGQQGKQGP   KGDVGVSGEQ    GVPGPPGPQG   QRGYPGMAGP   KGETGPAGYK
251  GMVGTIGAAG   RPGREGPKGP    PGDPGEKGEL   GGRGIRGPQG   DIGPKGDMGL
301  PGIDGKDGTP   GIPGVKGTAG    QPGRPGPPGH   RGQAGLPGQP   GSKGGPGDKG
351  EVGARGQQGI   TGTPGLDGEP    GPPGDAGTAG   VPGLKGDRGE   RGPVGAPGEA
401  GQSGPKGEQG   PPGIPGPQGL    PGVKGDKGSP   GKTGPKGSTG   DPGVHGLAGV
451  KGEKGESGEP   GPKGQQGIQG    ELGFPGPSGD   AGSPGVRGYP   GPPGPRGLLG
501  ERGVPGMPGQ   RGVAGRDAGD    QHIIDVVLKM   MQEQLAEVAV   SAKRAALGGV
551  GAMGPPGPPG   PPGPPGEQGL    HGPMGPRGVP   GLLGAAGQIG   NIGPKGKRGE
601  KGERGDTGRG   HPGMPGPPGI    PGLPGIPGHA   LAGKDGERGP   PGVPGDAGRP
651  GSPGPAGLPG   FCEPAACLGA    LPTPRHG
```

Structural and functional sites
Signal peptide: 1–21
NC4 domain: 22–24
COL3 domain: 25–161
NC3 domain: 162–178
COL2 domain: 179–517
NC2 domain: 518–547
COL1 domain: 548–662
NC1 domain: 663–677
GAG attachment site: 167
Lysine/hydroxylysine cross-linking site: 181 (from data on bovine type IX collagen)
Interchain disulphide bond residues: 172, 175, 662, 667
Imperfections in Gly–X–Y triplets: 112–116, 608–612, 618–622
Stromelysin cleavage site: 541–542

Accession number
S20819; S22429; S22918

Primary structure: α3(IX) chain (chicken)

Sequence conflicts: 196 G to E
406 S to A

Ala	A	42	Cys C 6	Asp D 29	Glu E 26
Phe	F	9	Gly G 204	His H 6	Ile I 24
Lys	K	30	Leu L 37	Met M 9	Asn N 9
Pro	P	138	Gln Q 21	Arg R 28	Ser S 22
Thr	T	20	Val V 14	Trp W 0	Tyr Y 1

Mol. wt (calc.) = 62 943   Residues = 675

```
  1    MTVFPTLGLL   FLCQLLATTS   AQRVGPQGPP   GPRGPPGPSG   KDGIDGEPGP
 51    SGLPGPPGPK   GAPGKPGAAG   EAGLPGLPGV   DGLTGTDGPP   GPNGPPGDRG
101    ALGPAGPPGP   AGKGLPGPPG   PPGPSGLPGG   NGFRGPPGPS   GLPGFPGPPG
151    PPGPPGLAGI   IPEGGGDLQC   PALCPPGPPG   PPGMPGFKGH   TGHKGGPGEI
201    GKEGEKGSPG   PPGPPGIPGS   VGLQGPRGLR   GLPGPMGPAG   DRGDIGFRGP
251    PGIPGPPGRA   GDQGNKGPQG   FRGPKGDTGR   PGPKGNPGAR   GLIGEPGIPG
301    KDGRDGAPGL   DGEKGDAARM   GVPGEKGPNG   LPGLPGRAGI   KGSKGEPGSP
351    GEMGEAGPSG   EPGIPGDVGI   PGDRGLPGPR   GATGPVGLPG   PIGAPGVRGF
401    QGPKGSSGEP   GLPGPTGIRG   ESGDRGPAGV   IGAKGSQGIA   GADGLPGDKG
451    ELGPFGPPGQ   KGEPGKRGEL   GPKGAQGPNG   TAGAPGIPGH   PGPMGHQGEQ
501    GVPGITGKPG   PPGKEASEQH   IRELCGEMIN   DQIAQLAANL   RKPLSPGMTG
551    RPGPAGPPGP   PGATGSVGHP   GARGPPGYRG   PTGELGDPGP   RGDTGEKGDK
601    GPAGQGIDGP   DGDQGPQGLP   GVPGISKNGR   DGAQGEPGLP   GDPGTPGAVG
651    AQGTPGICDT   SACMGAVGAS   TSKKS
```

Structural and functional sites
Signal peptide: 1–21
NC4 domain: 22–24
COL3 domain: 25–161
NC3 domain: 162–176
COL2 domain: 177–515
NC2 domain: 516–546
COL1 domain: 547–658
NC1 domain: 659–675
Cross-linking sites to α1(IX) and α3(IX) chains: 673, 674 (from data on bovine type IX collagen)
Lysine/hydroxylysine cross-linking sites: 188, 326 (from data on bovine type IX collagen)
Potential N-linked glycosylation site: 479
Interchain disulphide bond residues: 170, 174, 525, 658, 663
Imperfections in Gly–X–Y triplets: 112–113, 604–605, 624–628
Stromelysin cleavage site: 539–540

Gene structure

The use of the alternative promoter in the intron between exons 6 and 7 of the α1(IX) gene causes the loss of polypeptide encoded by exons 1–6. The new exon 1* encodes a different amino acid sequence in the shortened NC4 domain of the α1(IX) chain. The α1(IX) chain is encoded by a single gene found on human chromosome 6 at locus q13. The gene contains 19 exons and spans 100 kb. The chicken α2(IX) gene has 32 exons spanning approximately 10 kb [5-7]. The gene encoding the human α2 (IX) chain has been assigned to chromosome 1 [15].

References

1 Eyre, D.R. et al (1987) Collagen type IX: Evidence for covalent linkages to type II collagen in cartilage. FEBS Lett. 220: 337–341.

2 van der Rest, M. and Mayne, R. (1987) Type IX collagen. In: Structure and Function of Collagen Types, Mayne, R. and Burgeson, R.E., eds, Academic Press, London, pp. 195–221.

3 van der Rest, M. and Mayne, R. (1988) Type IX collagen from cartilage is covalently crosslinked to type II collagen. J. Biol. Chem. 263: 1615–1618.

4 Kimura, T. et al (1989) The complete primary structure of two distinct forms of human α1(IX) collagen chains. Eur. J. Biochem. 179: 71–78.

5 Nishimura, I. et al (1989) Tissue-specific forms of type IX collagen-proteoglycan arise from the use of two widely separated promoters. J. Biol. Chem. 264: 20033–20041.

6 Muragaki, Y. et al (1990) Molecular cloning of rat and human type IX collagen cDNA and localization of the α1(IX) gene on the human chromosome 6. Eur. J. Biochem. 192: 703–708.

7 Ninomiya, Y. et al (1990) The molecular biology of collagens with short triple-helical domains. In: Extracellular Matrix Genes, Sandell, L.J. and Boyd, C.D., eds, Academic Press, New York, pp. 79–114.

8 Ayad, S. et al (1991) Mammalian cartilage synthesizes both proteoglycan and non-proteoglycan forms of type IX collagen. Biochem. J. 278: 441–445.

9 Brewton, R.G. et al (1992) Cloning of the chicken α3(IX) collagen chain completes the primary structure of type IX collagen. Eur. J. Biochem. 205: 443–449.

10 Har-El, R. et al (1992) Cloning and developmental expression of the α3(IX) chain of chicken type IX collagen. J. Biol. Chem. 267: 10070–10076.

11 Wu, J.-J. et al (1992) Identification of cross-linking sites in bovine cartilage type IX collagen reveals an antiparallel type II-type IX molecular relationship and type IX to type IX bonding. J. Biol. Chem. 267: 23007–23014.

12 Duance, V.C. et al (1984) Isolation and characterization of the precursor of type M collagen. Biochem. J. 221: 885–889.

13 Ayad, S. et al (1989) Bovine cartilage types VI and IX collagens. Characterization of their forms in vivo. Biochem. J. 262: 753–761.

14 Grant, M.E. et al (1988) The structure and synthesis of cartilage collagens. In: The Control of Tissue Damage, Glauert, A.M., ed., Elsevier, Amsterdam, pp. 3–28.

15 Perälä, M. et al (1993) Molecular Cloning of the human α2(1X)) collagen cDNA and assignment of the human COL9A2 gene to chromosome 1. FEBS Lett. 319: 177–180.

Collagen type X

Type X collagen is a short-chain collagen which is both temporally and spatially regulated during foetal development. It is synthesized predominantly by hypertrophic chondrocytes during endochondral bone formation and therefore has a very restricted tissue distribution in the calcifying cartilage that is eventually replaced by bone.

Molecular structure

Type X collagen is a homotrimer comprising three identical α1(X) chains. It has a short triple helix approximately 132 nm in length, a small NH$_2$-terminal domain and a large COOH-terminal globular domain. It is deposited in the cartilage matrix without apparent processing and, although its macromolecular organization has not been determined *in vivo*, it may form a hexagonal type lattice as in the case of type VIII collagen. The COOH-terminal three-quarters of the NC1 domain is homologous to that of the two type VIII collagen chains. The lengths of the triple-helical and NC1 domains are similar to those of type VIII collagen, but the NC2 domain is shorter [1–5].

Isolation

Type X collagen is isolated in its intact form from the medium of cultured chondrocytes [6] or as its triple-helical domain by pepsinization [7].

Accession number
S18249

Primary structure: α1(X) chain

Ala	A	36	Cys	C	1	Asp	D	12	Glu	E	22

Ala A 36   Cys C 1   Asp D 12   Glu E 22
Phe F 15   Gly G 175   His H 10   Ile I 27
Lys K 35   Leu L 35   Met M 10   Asn N 14
Pro P 145   Gln Q 23   Arg R 19   Ser S 27
Thr T 25   Val V 26   Trp W 2   Tyr Y 21
Mol. wt (calc.) = 66 053   Residues = 680

```
1     MLPQIPFLLL   VSLNLVHGVF   YAERYQTPTG   IKGPLPNTKT   QFFIPYTIKS
51    KGIAVRGEQG   TPGPPGPAGP   RGHPGPSGPP   GKPGYGSPGL   QGEPGLPGPP
101   GPSAVGKPGV   PGLPGKPGER   GPYGPKGDVG   PAGLPGPRGP   PGPPGIPGPA
151   GISVPGKPGQ   QGPTGAPGPR   GFPGEKGAPG   VPGMNGQKGE   MGYGAPGRPG
201   ERGLPGPQGP   TGPSGPPGVG   KRGENGVPGQ   PGIKGDRGFP   GEMGPIGPPG
251   PQGPPGERGP   EGIGKPGAAG   APGQPGIPGT   KGLPGAPGIA   GPPGPPGFGK
301   PGLPGLKGER   GPAGLPGGPG   AKGEQGPAGL   PGKPGLTGPP   GNMGPQGPKG
351   IPGSHGLPGP   KGETGPAGPA   GYPGAKGERG   SPGSDGKPGY   PGKPGLDGPK
401   GNPGLPGPKG   DPGVGGPPGL   PGPVGPAGAK   GMPGHNGEAG   PRGAPGIPGT
451   RGPIGPPGIP   GFPGSKGDPG   SPGPPGPAGI   ATKGLNGPTG   PPGPPGPRGH
```

```
501   SGEPGLPGPP   GPPGPPGQAV   MPEGFIKAGQ   RPSLSGTPLV   SANQGVTGMP
551   VSAFTVILSK   AYPAIGTPIP   FDKILYNRQQ   HYDPRTGIFT   CQIPGIYYFS
601   YHVHVKGTHV   WVGLYKNGTP   VMYTYDEYTK   GYLDQASGSA   IIDLTENDQV
651   WLQLPNAESN   GLYSSEYVHS   SFSGFLVAPM
```

Structural and functional sites
Signal peptide: 1–18
NC2 domain: 19–56
Helical domain: 57–519
NC1 domain: 520–680
Potential N-linked glycosylation site: 617
Imperfections in Gly–X–Y triplets: 84–85, 101–105, 151–155, 192–193, 218–219, 222–223, 297–298, 479–480 (same relative locations as imperfections in α1(VIII) and α2(VIII))
Mammalian collagenase cleavage sites: 151–152, 479–480

Gene structure

The human type X collagen gene spans approximately 7 kb and is located on the long arm of chromosome 6 at locus q21–22.3. Its structure is unique in comparison with other known vertebrate collagen genes in that it is condensed, comprising only three exons. Exon 3 encodes $4^2/_3$ amino acids of the NH_2-terminal domain, the entire triple helix, the COOH-terminal globular domain and part of the 3'-untranslated region [3-5].

References
[1] Schmid, T.M. et al (1986) Type X collagen contains two cleavage sites for a vertebrate collagenase. J. Biol. Chem. 261: 4184–4189.
[2] Schmid, T.M. and Linsenmayer, T.F. (1987) Type X collagen. In: Structure and Function of Collagen Types, Mayne, R. and Burgeson, R.E., eds, Academic Press, London, pp. 223–259.
[3] La Valle, P. et al (1988) The type X collagen gene intron sequences split the 5'-untranslated region and separate the coding regions for the non-collagenous amino-terminal and triple helical domains. J. Biol. Chem. 263: 18378–18385.
[4] Apte, S. et al (1991) Cloning of human α1(X) collagen DNA and localization of the COL10A1 gene to the q21–q22 region of human chromosome 6. FEBS Lett. 282: 393–396.
[5] Thomas, J.T. et al (1991) The human collagen X collagen. Complete primary translated sequence and chromosomal localization. Biochem. J. 280: 617–623.
[6] Schmid, T.M. and Linsenmayer, T.F. (1983) A short chain (pro)collagen from aged endochondral chondrocytes. Biochemical characterization. J. Biol. Chem. 258: 9504–9509.
[7] Kielty, C.M. et al (1984) Embryonic chick cartilage collagens. Differences in the low-Mr species present in sternal cartilage and tibiotarsal articular cartilage. FEBS Lett. 169: 179–184.

Collagen type XI

Type XI collagen is a quantitatively minor fibrillar collagen and forms heterotypic fibrils with types II and IX collagens in cartilaginous tissues.

Molecular structure

Type XI collagen is synthesized as a heterotrimeric procollagen comprising three distinct proα1(XI), proα2(XI) and proα3(XI) chains, although alternative assemblies may exist. The proα1(V) chain can substitute for the proα1(XI) chain forming cross-type heterotrimers. The triple-helical domain of the proα3(XI) chain is probably a post-translationally modified (more glycosylated) form of the proα1(II) chain. However, the proα3(XI) chain is not cleaved by N-proteinase or mammalian collagenase, suggesting differences between the two chains either in amino acid sequence or enzyme accessibility. The procollagen forms are processed extracellularly at the COOH-terminus, but only partial processing occurs at the NH2-terminus. Type XI collagen is co-polymerized in the interior of type II collagen fibrils [1–4].

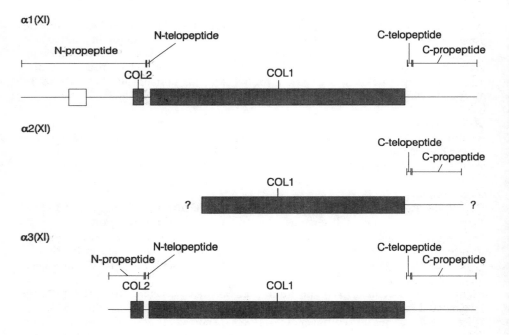

Isolation

Type XI collagen can be prepared from cartilage as its shorter triple-helical form by guanidine–HCl treatment (to remove proteoglycans) and pepsin digestion [5].

Accession number

P12107

Primary structure: α1(XI) chain

Ala	A	97	Cys	C	12	Asp	D	97	Glu	E	122
Phe	F	47	Gly	G	423	His	H	16	Ile	I	56
Lys	K	105	Leu	L	80	Met	M	24	Asn	N	36
Pro	P	286	Gln	Q	81	Arg	R	70	Ser	S	73
Thr	T	70	Val	V	63	Trp	W	10	Tyr	Y	36

Mol. wt (calc.) = 180 718   Residues = 1806

```
1     MEPWSSRWKT  KRWLWDFTVT  TLALTFLFQA  REVRGAAPVD  VLKALDFHNS
51    PEGISKTTGF  CTNRKNSKGS  DTAYRVSKQA  QLSAPTKQLF  PGGTFPEDFS
101   ILFTVKPKKG  IQSFLLSIYN  EHGIQQIGVE  VGRSPVFLFE  DHTGKPAPED
151   YPLFRTVNIA  DGKWHRVAIS  VEKKTVTMIV  DCKKKTTKPL  DRSERAIVDT
201   NGITVFGTRI  LDEEVFEGDI  QQFLITGDPK  AAYDYCEHYS  PDCDSSAPKA
251   AQAQEPQIDE  YAPEDIIEYD  YEYGEAEYKE  AESVTEGPTV  TEETIAQTEA
301   NIVDDFQEYN  YGTMESYQTE  APRHVSGTNE  PNPVEEIFTE  EYLTGEDYDS
351   QRKNSEDTLY  ENKEIDGRDS  DLLVDGDLGE  YDFYEYKEYE  DKPTSPPNEE
401   FGPGVPAETD  ITETSINGHG  AYGEKGQKGE  PAVVEPGMLV  EGPPGPAGPA
451   GIMGPPGLQG  PTGPPGDPGD  RGPPGRPGLP  GADGLPGPPG  TMLMLPFRYG
501   GDGSKGPTIS  AQEAQAQAIL  QQARIALRGP  PGPMGLTGRP  GPVGGPGSSG
551   AKGESGDPGP  QGPRGVQGPP  GPTGKPGKRG  RPGADGGRGM  PGEPGAKGDR
601   GFDGLPGLPG  DKGHRGERGP  QGPPGPPGDD  GMRGEDGEIG  PRGLPGEAGP
651   RGLLGPRGTP  GAPGQPGMAG  VDGPPGPKGN  MGPQGEPGPP  GQQGNPGPQG
701   LPGPQGPIGP  PGEKGPQGKP  GLAGLPGADG  PPGHPGKEGQ  SGEKGALGPP
751   GPQGPIGXPG  PRGVKGADGV  RGLKGSKGEK  GEDGFPGFKG  DMGLKGDRGE
801   VGQIGPRGXD  GPEGPKGRAG  PTGDPGPSGQ  AGEKGKLGVP  GLPGYPGRQG
851   PKGSTGFPGF  PGANGEKGAR  GVAGKPGPRG  QRGPTGPRGS  RGARGPTGKP
901   GPKGTSGGDG  PPGPPGERGP  QGPQGPVGFP  GPKGPPGPPG  RMGCPGHPGQ
951   RGETGFQGKT  GPPGPGGVVG  PQGPTGETGP  IGERGYPGPP  GPPGEQGLPG
1001  AAGKEGAKGD  PGPQGISGKD  GPAGLRGFPG  ERGLPGAQGA  PGLKGGEGPQ
1051  GPPGPVGSPG  ERGSAGTAGP  IGLRGRPGPQ  GPPGPAGEKG  APGEKGPQGP
1101  AGRDGVQGPV  GLPGPAGPAG  SPGEDGDKGE  IGEPGQKGSK  GGKGENGPPG
1151  PPGLQGPVGA  PGIAGGDGEP  GPRGQQGMFG  QKGDEGARGF  PGPPGPIGLQ
1201  GLPGPPGEKG  ENGDVGPWGP  PGPPGPRGPQ  GPNGADGPQG  PPGSVGSVGG
1251  VGEKGEPGEA  GNPGPPGEAG  VGGPKGERGE  KGEAGPPGAA  GPPGAKGPPG
1301  DDGPKGNPGP  VGFPGDPGPP  GELGPAGQDG  VGGDKGEDGD  PGQPGPPGPS
1351  GEAGPPGPPG  KRGPPGAAGA  EGRQGEKGAK  GEAGAEGPPG  KTGPVGPQGP
1401  AGKPGPEGLR  GIPGPVGEQG  LPGAAGQDGP  PGPMGPPGLP  GLKGDPGSKG
1451  EKGHPGLIGL  IGPPGEQGEK  GDRGLPGTQG  SPGAKGDGGI  PGPAGPLGPP
1501  GPPGLPGPQG  PKGNKGSTGP  AGQKGDSGLP  GPPGPPGPPG  EVIQPLPILS
1551  SKKTRRHTEG  MQADADDNIL  DYSDGMEEIF  GSLNSLKQDI  EHMKFPMGTQ
1601  TNPARTCKDL  QLSHPDFPDG  EYWIDPNQGC  SGDSFKVYCN  FTSGGETCIY
1651  PDKKSEGVRI  SSWPKEKPGS  WFSEFKRGKL  LSYLDVEGNS  INMVQMTFLK
1701  LLTASARQNF  TYHCHQSAAW  YDVSSGSYDK  ALRFLGSNDE  EMSYDNNPFI
1751  KTLYDGCTSR  KGYEKTVIEI  NTPKIDQVPI  VDVMISDFGD  QNQKFGFEVG
1801  PVCFLG
```

Structural and functional sites

Signal peptide: 1–36 (probable)
N-Propeptide: 37–528
PARP repeat 37–259

COL2 domain: 420–511 (interrupted triple helix of 4, 17 and 3 triplets similar to the COL2 domain in proα2(V), and proα1(V))
Helical domain: 529–1542
C-Telopeptide: 1543–1563
C-Propeptide: 1564–1806
Lysine/hydroxylysine cross-linking sites: 612, 1452
Potential N-linked glycosylation sites: 1640, 1709
C-Proteinase cleavage site: 1563–1564

Primary structure: α2(XI) chain

Only 80% of the main triple-helical domain and part of the C-propeptide have been sequenced (Accession: P13942) [3].

Primary structure: α3(XI) chain

The main triple-helical domain is apparently identical to α1(II), but is more glycosylated on hydroxylysine residues.

Gene structure

The proα1(XI) and proα2(XI) collagen chains are encoded by single genes found on human chromosomes 1 (locus p21) and 6 (locus p21.2), respectively. The proα3(XI) chain is probably encoded by the same gene as the proα1(II) chain (chromosome 12, locus q13.11–12) but is modified post-translationally [3].

References
[1] Eyre, D.R. and Wu, J.-J. (1987) Type XI or 1α2α3α collagen. In: Structure and Function of Collagen Types, Mayne, R. and Burgeson, R.E., eds, Academic Press, London, pp. 261–281.
[2] Bernard, M. et al (1988) Cloning and sequencing of pro-α1(XI) collagen cDNA demonstrates that type XI belongs to the fibrillar class of collagens and reveals that the expression of the gene is not restricted to cartilagenous tissue. J. Biol. Chem. 263: 17159–17166.
[3] Kimura, T. et al (1989) The human α2(XI) chain. Molecular cloning of cDNA and genomic DNA reveals characteristics of a fibrillar collagen with differences in genomic organisation. J. Biol. Chem. 264: 13910–13916.
[4] Yoshioka, H. and Ramirez, F. (1990) Pro-α1(XI) collagen. Structure of the amino-terminal propeptide and expression of the gene in tumor cell lines. J. Biol. Chem. 265: 6423–6426.
[5] Grant, M.E. et al (1988) The structure and synthesis of cartilage collagens. In: The Control of Tissue Damage, Glauert, A.M., ed., Elsevier, Amsterdam, pp. 3–28.

Collagen type XII

Type XII collagen is a member of the FACIT collagens that include collagen types IX, XIV and XVI. Type XII collagen is found mainly in tissues rich in type I collagen. However, recent studies indicate that it is also present in cartilaginous tissues that contain type II collagen.

Molecular structure

Type XII collagen is synthesized as a disulphide-bonded homotrimeric molecule comprising three α1(XII) chains. The molecule is not processed prior to deposition in the extracellular matrix. Each chain consists of two collagenous triple-helical domains (COL1–2) and three non-collagenous domains (NC1–3). The COL1 domain is homologous to the COL1 domains of type IX, XIV and XVI collagens. Type XII collagen does not possess a domain homologous to COL2(IX) and therefore is not cross-linked either to itself or to other collagens by lysine/hydroxylysine-derived bonds. Type XII collagen variants, differing in the size of their NC3 domains, arise by alternative splicing. The NC3 domain of the long form is composed of 18 fibronectin type III repeats, four von Willebrand factor A repeats and a PARP repeat. In some tissues, e.g. foetal bovine cartilage, the larger variant may possess a glycosaminoglycan chain attached to the NH2-terminal portion of the NC3 domain and is therefore a proteoglycan [1-7].

Isolation
Type XII collagen variants are readily extracted from tissues by low-molarity NaCl solutions since they are not covalently cross-linked within the extracellular matrix [1,5,7].

Accession number
A40020

Primary structure: α1(XII) chain (chicken)

Ala A 174	Cys C 24	Asp D 172	Glu E 201
Phe F 91	Gly G 283	His H 28	Ile I 155
Lys K 136	Leu L 207	Met M 47	Asn N 133
Pro P 270	Gln Q 103	Arg R 181	Ser S 232
Thr T 254	Val V 282	Trp W 30	Tyr Y 121

Mol. wt (calc.) = 340 199   Residues = 3124

```
1     MRTALCSAVA  ALCAAALLSS  IEAEVNPPSD  LNFTIIDEHN  VQMSWKRPPD
51    AIVGYRITVV  PTNDGPTKEF  TLSPSTTQTV  LSDLIPEIEY  VVSIASYDEV
101   EESLPVFGQL  TIQTGGPGIP  EEKKVEAQIQ  KCSISAMTDL  VFLVDGSWSV
151   GRNNFRYILD  FMVALVSAFD  IGEEKTRVGV  VQYSSDTRTE  FNLNQYFRRS
201   DLLDAIKRIP  YKGGNTMTGE  AIDYLVKNTF  TESAGARKGF  PKVAIVITDG
251   KAQDEVEIPA  RELRNIGVEV  FSLGIKAADA  KELKLIASQP  SLKHVFNVAN
301   FDGIVDIQNE  IILQVCSGVD  EQLGELVSGE  EVVPEASNLV  ATQISSKSVR
```

351	ITWDPSTSQI	TGYRVQFIPM	IAGGKQHVLS	VGPQTTALNV	KDLSPDTEYQ
401	INVYAMKGLT	PSEPITIMEK	TQQVKVQVEC	SRGVDVKADV	VFLVDGSYSI
451	GIANFVKVRA	FLEVLVKSFE	ISPRKVQISL	VQYSRDPHME	FSLNRYNRVK
501	DIIQAINTFP	YRGGSTNTGK	AMTYVREKVF	VTSKGSRPNV	PRVMILITDG
551	KSSDAFKEPA	IKLRDADVEI	FAVGVKDAVR	TELEAIASPP	AETHVYTVED
601	FDAFQRISFE	LTQSVCLRIE	QELAAIRKKS	YVPAKNMVFS	DVTSDSFKVS
651	WSAAGSEEKS	YLIKYKVAIG	GDEFIVSVPA	SSTSSVLTNL	LPETTYAVSV
701	IAEYEDGDGP	PLDGEETTLE	VKGAPRNLRI	TDETTDSFIV	GWTPAPGNVL
751	RYRLVYRPLT	GGERRQVTVS	ANERSTTLRN	LIPDTRYEVS	VIAEYQSGPG
801	NALNGYAKTD	EVRGNPRNLR	VSDATTSTTM	KLSWSAAPGK	VQHVLYNLHT
851	RYAGVETKEL	TVKGDTTSKE	LKGLDEATRY	ALTVSALYAS	GAGEALSGEG
901	ETLEERGSPR	NLITTDITDT	TVGLSWTPAP	GTVNNYRIVW	KSLYDDTMGE
951	KRVPGNTVDA	VLDGLEPETK	YRISIYAAYS	SGEGDPVEGE	AFTDVSQSAR
1001	TVTVDNETEN	TMRVSVAALT	WEGLVLARVL	PNRSGGRQMF	GKVNASATSI
1051	VLKRLKPRTT	YDLSVVPIYD	FGQGKSRKAE	GTTASPFKPP	RNLRTSDSTM
1101	SSFRVTWEPA	PGRVKGYKVT	FHPTEDDRNL	GELVVGPYDS	TVVLEELRAG
1151	TTYKVNVFGM	FDGGESNPLV	GQEMTTLSDT	TTEPFLSRGL	ECRTRAEADI
1201	VLLVDGSWSI	GRPNFKTVRN	FISRIVEVFD	IGPDKVQIGL	AQYSGDPRTE
1251	WNLNAYRTKE	ALLDAVTNLP	YKGGNTLTGM	ALDFILKNNF	KQEAGLRPRA
1301	RKIGVLITDG	KSQDDVVTPS	RRLRDEGVEL	YAIGIKNADE	NELKQIATDP
1351	DDIHAYNVAD	FSFLASIGED	VTTNLCNSVK	GPGDLPPPSN	LVISEVTPHS
1401	FRLRWSPPPE	SVDRYRVEYY	PTTGGPPKQF	YVSRMETTTV	LKDLTPETEY
1451	IVNVFSVVED	ESSEPLIGRE	ITYPLSSVRN	LNVYDIGSTS	MRVRWEPVNG
1501	ATGYLLTYEP	VNATVPTTEK	EMRVGPSVNE	VQLVDLIPNT	EYTLTAYVLY
1551	GDITSDPLTS	QEVTLPLPGP	RGVTIRDVTH	STMNVLWDPA	PGKVRKYIIR
1601	YKIADEADVK	EVEIDRLKTS	TTLTDLSSQR	LYNVKVVAVY	DEGESLPVVA
1651	SCYSAVPSPV	NLRITEITKN	SFRGTWDHGA	PDVSLYRITW	GPYGRSEKAE
1701	SIVNGDVNSL	LFENLNPDTL	YEVSVTAIYP	DESETVDDLI	GSERTLPLVP
1751	ITTPAPKSGP	RNLQVYNATS	HSLTVKWDPA	SGRVQRYKII	YQPINGDGPE
1801	QSTMVGGRQN	SVVIQKLQPD	TPYAITVSSM	YADGEGGRMT	GRGRTKPLTT
1851	VKNMLVYDPT	TSTLNVRWDH	AEGNPRQYKV	FYRPTAGGAE	EMTTVPGNTN
1901	YVILRSLEPN	TPYTVTVVPV	FPEGDGGRTT	DTGRTLERGT	PRNIQVYNPT
1951	PNSMNVRWEP	APGPVQQYRV	NYSPLSGPRP	SESIVVPANT	RDVMLERLTP
2001	DTAYSINVIA	LYADGEGNPS	QAQGRTLPRS	GPRNLRVFDE	TTNSLSVQWD
2051	HADGPVQQYR	IIYSPTVGDP	IDEYTTVPGI	RNNVILQPLQ	SDTPYKITVV
2101	AVYEDGDGGQ	LTGNGRTVGL	LPPQNIYITD	EWYTRFRVSW	DPSPSPVLGY
2151	KIVYKPVGSN	EPMEVFVGEV	TSYTLHNLSP	STTYDVNVYA	QYDSGMSIPL
2201	TDQGTTLYLN	VTDLTTYKIG	WDTFCIRWSP	HRSATSYRLK	LNPADGSRGQ
2251	EITVRGSETS	HCFTGLSPDT	EYNATVFVQT	PNLEGPPVSV	REHTVLKPTE
2301	APTPPPTPPP	PPTIPPARDV	CRGAKADIVF	LTDASWSIGD	DNFNKVVKFV
2351	FNTVGAFDLI	NPAGIQVSLV	QYSDEAQSEF	KLNTFDDKAQ	ALGALQNVQY
2401	RGGNTRTGKA	LTFIKEKVLT	WESGMRRGVP	KVLVVVTDGR	SQDEVRKAAT
2451	VIQHSGFSVF	VVGVADVDYN	ELAKIASKPS	ERHVFIVDDF	DAFEKIQDNL
2501	VTFVCETATS	TCPLIYLEGY	TSPGFKMLES	YNLTEKHFAS	VQGVSLESGS
2551	FPSYVAYRLH	KNAFVSQPIR	EIHPEGLPQA	YTIIMLFRLL	PESPSEPFAI
2601	WQITDRDYKP	QVGVVLDPGS	KVLSFFNKDT	RGEVQTVTFD	NDEVKKIFYG
2651	SFHKVHIVVT	SSNVKIYIDC	SEILEKPIKE	AGNITTDGYE	ILGKLLKGDR
2701	RSATLEIQNF	DIVCSPVWTS	RDRCCDLPSM	RDEAKCPALP	NACTCTQDSV
2751	GPPGPPGPPG	GPGAKGPRGE	RGLTGSSGPP	GPRGETGPPG	PQGPPGPQGP
2801	NGLQIPGEPG	RQGMKGDAGQ	PGLPGRSGTP	GLPGPPGPVG	PPGERGFTGK

```
2851   DGPTGPRGPP   GPAGAPGVPG   VAGPSGKPGK   PGDRGTPGTP   GMKGEKGDRG
2901   DIASQNMMRA   VARQVCEQLI   NGQMSRFNQM   LNQIPNDYYS   NRNQPGPPGP
2951   PGPPGAAGTR   GEPGPGGRPG   FPGPPGVQGP   PGERGMPGEK   GERGTGSQGP
3001   RGLPGPPGPQ   GESRTGPPGS   TGSRGPPGPP   GRPGNAGIRG   PPGPPGYCDS
3051   SQCASIPYNG   QGFPEPYVPE   SGPYQPEGEP   FIVPMESERR   EDEYEDYGVE
3101   MHSPEYPEHM   RWKRSLSRKA   KRKP
```

Structural and functional sites

Signal peptide: 1–23

NC3 domain: 24–2750

Alternatively spliced domain: 25–1188

COL2 domain: 2751–2902

NC2 domain: 2903–2945

COL1 domain: 2946–3048

NC1 domain: 3049–3124

Fibronectin type III repeats: 24–114, 332–425, 629–720, 721–811, 812–904, 905–998, 999–1085, 1086–1178, 1384–1473, 1474–1565, 1566–1654, 1655–1755, 1756–1846, 1847–1936, 1937–2027, 2028–2118, 2119–2207, 2208–2295

von Willebrand factor A repeats: 128–319, 426–619, 1188–1379, 2297–2508

PARP repeat: 2509–2750

Potential O-linked glycosylation sites: 1389, 1397, 1406, 2299, 2303, 2307, 2313

Potential N-linked glycosylation sites: 32, 1006, 1032, 1044, 1512, 1767, 1948, 1971, 2018, 2177, 2210, 2273, 2532, 2683

Interchain disulphide bonds: 3048, 3053 (conserved in type IX collagen)

Imperfections in Gly–X–Y triplets: 2994–2995, 3014–3015 (conserved in the COL1 domain of all three chains of type IX collagen), 2805–2806

Gene structure

The gene for the α1(XII) chain has been assigned to chromosome 6 at locus q12–14. This is the same locus as the α1(IX) chain gene.

References

[1] Dublet, B. et al (1989) The structure of avian type XII collagen. α1(XII) chains contain 190-kDa non-triple helical amino terminal domains and form homotrimeric molecules. J. Biol. Chem. 264: 13150–13156.

[2] Gordon, M.K. et al (1989) Type XII collagen. A large multidomain molecule with partial homology to type IX collagen. J. Biol. Chem. 264: 19772–19778.

[3] Yamagata, M. et al (1991) The complete primary structure of type XII collagen shows a chimeric molecule with reiterated fibronectin type III motifs, von Willebrand factor A motifs, a domain homologous to a noncollagenous region of type IX collagen, and short collagenous domains with an arg-gly-asp site. J. Cell Biol. 115: 209–221.

[4] Koch, M. et al (1992) A major oligomeric fibroblast proteoglycan identified as a novel large form of type XII collagen. Eur. J. Biochem. 207: 847–856.

[5] Lunstrum, G.P. et al (1992) Identification and partial characterization of a large variant form of type XII collagen. J. Biol. Chem. 267: 20087–20092.

6 Trueb, J. and Trueb, B. (1992) The splice variants of collagen XII share a common 5′ end. Biochim. Biophys. Acta 1171: 97–98.

7 Watt, S.L. et al (1992) Characterization of collagen types XII and XIV from fetal bovine cartilage. J. Biol. Chem. 267: 20093–20099.

Collagen type XIII

Type XIII is a short-chain, non-fibrillar collagen that is expressed in skin, intestine, placenta, bone, cartilage and striated muscle. It has, however, only been characterized at the genomic and cDNA levels and the protein has yet to be isolated.

Molecular structure

Type XIII is probably synthesized as a homotrimer of three α1(XIII) chains, but whether the three chains exhibit identical splicing is not known. Each chain consists of three collagenous domains (COL1–3) and four non-collagenous domains (NC1–4). Complex alternative splicing of the primary transcript occurs in both collagenous and non-collagenous domains. Alternatively spliced forms may vary between 516 and 623 amino acids and are generated by exon skipping (exons 3B, 4A, 4B and 5 in COL3; exons 12 and 13 in NC3; exons 29 and 33 in COL1 and exon 37 at the COL1/NC1 junction site). At least 12 mRNA species exist through the alternations of exons 3B-5, 12 and 13, and distinct differences in the proportions of the variant mRNAs were observed in four cultured cell lines and seven tissues. Furthermore, four combinations of alternatively spliced exons 29–37 have been found among cDNA clones isolated from human endothelial cells and HT1080 cells, but additional combinations are predicted to occur.

N.B. While these COL and NC domains designations follow the usual convention and are numbered from C to N, it should be noted that in Pihlajaniemi and Tamminen [1], the domains are numbered from N to C [1–4].

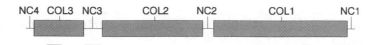

Isolation
Type XIII collagen protein has not been isolated.

Accession number
A38298; B38298; C38298

Primary structure: α1(XIII) chain

Ala	A	29	Cys	C	5	Asp	D	23	Glu	E	36
Phe	F	5	Gly	G	178	His	H	9	Ile	I	17
Lys	K	45	Leu	L	38	Met	M	11	Asn	N	11
Pro	P	119	Gln	Q	25	Arg	R	24	Ser	S	16
Thr	T	13	Val	V	14	Trp	W	2	Tyr	Y	3

Mol. wt (calc.) = 60 104   Residues = 623

```
1     METAILGRVN   QLLDEKWKLH    SRRRREAPKT   SPGCNCPPGP   PGPTGRPGLP
51    GDKGAIGMPG   RVGVKGQPGE    KGSPGDAGLS   IIGPRGPPGQ   PGTRGFPGFP
101   GPIGLDGFPG   HPGPKGDMGL    TGPPGQPGPQ   GQKGEKGQCG   EYPHRLLPLL
151   NSVRLAPPPV   IKRRTFQGEQ    SQASIQGPPG   PPGPPGPSGP   LGHPGLPGPM
```

```
201   GPPGLPGPPG   PKGDPGIQGY   HGRKGERGMP   GMPGKHGAKG   APGIAVAGMK
251   GEPGIPGTKG   EKGAEGSPGL   PGLLGQKGEK   GDAGNSIGGG   RGEPGPPGLP
301   GPPGPKGEAG   VDGQVGPPGQ   PGDKGERGAA   GEQGPDGPKG   SKGEPGKGEM
351   VDYNGNINEA   LQEIRTLALM   GPPGLPGQIG   PPGAPGIPGQ   KGEIGLLGPL
401   GHDGKGPRGK   PGDMGPPGPQ   GPPGKDGPPG   VKGENGHPGS   PGEKGEKGET
451   GQAGSPGEKG   EAGEKGNPGA   EVPGLPGPEG   PPGPPGLQGV   PGPKGEAGLD
501   GAKGEKGFQG   EKGDRGPLGL   PGASGLDGRP   GPPGTPGPIG   VPGPAGPKGE
551   RGSKGDPGMT   GPTGAAGLPG   LHGPPGDKGN   RGERGKKGST   GPKGDKGDQG
601   APGLDAPCPL   GEDGLPVQGC   WNK
```

Structural and functional sites
Signal peptide: 1–21
NC4 domain: 22–38
COL3 domain: 39–142
NC3 domain: 143–176
COL2 domain: 177–348
NC2 domain: 349–370
COL1 domain: 371–605
NC1 domain: 606–623
Alternatively spliced domains: 42–88, 146–167 (replaced by ECLSSMPAALRSSQIIALK; this gives a sequence that is three residues shorter), 457–471, 522–533, 582–610

Gene structure
The gene encoding the α1(XIII) chain spans at least 140 kb and comprises 40/41 exons. Twenty-one exons are multiples of 9 bp, the remaining exons vary between 24 and 153 bp. The gene has been localized to chromosome 10 at locus q22 [2,5].

References
[1] Pihlajaniemi, T. and Tamminen, M. (1990) The α1 chain of type XIII collagen consists of three collagenous and four noncollagenous domains, and its primary transcript undergoes complex alternative splicing. J. Biol. Chem. 265: 16922–16928.
[2] Tikka, L. et al (1991) Human α1(XIII) collagen gene. Multiple forms of the gene transcripts are generated through complex alternative splicing of several short exons. J. Biol. Chem. 266: 17713–17719.
[3] Juvonen, M. and Pihlajaniemi, T. (1992) Characterisation of the spectrum of alternative splicing of α1(XIII) collagen transcripts in HT1080 cells and calvarial tissue resulted in identification of two previously unidentified alternatively spliced sequences, one previously unidentified exon, and nine new mRNA variants. J. Biol. Chem. 267: 24693–24699.
[4] Juvonen, M. et al (1992) Patterns of expression of the six alternatively spliced exons affecting the structures of the COL1 and NC2 domains of the α1(XIII) collagen chain in human tissues and cell lines. J. Biol. Chem. 267: 24700–24707.
[5] Shows, T.B. et al (1989) Assignment of the human collagen α1(XIII) chain gene (COL13A1) to the q22 region of chromosome 10. Genomics 5: 128–133.

Collagen type XIV

Type XIV collagen is a member of the FACIT collagens that include types IX, XII and XVI collagen. Type XIV collagen is found mainly in tissues rich in collagen type I. However, recent studies indicate that it is also present in cartilaginous tissues that contain type II collagen.

Molecular structure

Type XIV collagen is synthesized as a disulphide-bonded homotrimeric molecule comprising three α1(XIV) chains and is not processed prior to deposition in the extracellular matrix. Each chain consists of two collagenous triple-helical domains (COL1–2) and three non-collagenous domains (NC1–3). The COL1 domain of α1(XIV) is homologous to the COL1 domain of type IX, XII and XVI collagens. The NC3 domain comprises eight fibronectin type III repeats, two von Willebrand factor A repeats and a PARP repeat. Type XIV collagen contains two cysteine residues that participate in interchain disulphide bonding (residues 1769 and 1774) and that are conserved in types IX and XII collagen. The triple-helical domain also contains two imperfections in Gly–X–Y triplets (residues 1715–1716 and 1735–1736) that are conserved in the COL1 domain of type IX and XII collagens. The seven fibronectin type III repeats (352–712, 741–1010) and the connecting region show 71% sequence identity with a non-collagenous protein, undulin. Type XIV collagen and undulin may arise through alternative splicing of a single pre-mRNA. In addition, two variants of type XIV exist which differ in their NC1 domains and which may arise by alternative splicing. As for types IX and XII collagen, type XIV collagen can also exist in a proteoglycan form in some tissues (e.g. foetal bovine cartilage) [1–8].

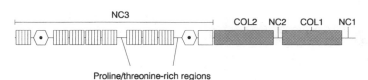

Proline/threonine-rich regions

Isolation

The intact, native form of type XIV collagen can be isolated by extraction with low-molarity NaCl solutions since it is not covalently cross-linked within the extracellular matrix [4,6].

Accession number

X70792 and X70793

Primary structure: α1(XIV) chain (chicken)

Sequence conflicts:	30–145 (absent in ref. 8)
	187 A to R
	248 S to C
	284 D to E

Ala	A	130	Cys	C	20	Asp	D	100	Glu	E	120
Phe	F	65	Gly	G	201	His	H	26	Ile	I	99

Lys	K	96	Leu	L	143	Met	M	27	Asn	N	61
Pro	P	166	Gln	Q	73	Arg	R	80	Ser	S	137
Thr	T	136	Val	V	131	Trp	W	16	Tyr	Y	61

Mol. wt (calc.) = 202 456   Residues = 1888

```
1     MLCWNEVQSC  FLLAFLAVAA  YSVSDAQGQV  SPPTRLRYNV  VNPDSVQISW
51    KAPKGQFSGY  KLLVTPSSGG  KTNQLILQNT  ATKAIIQGLI  PDQNYALQII
101   AFSDDKESKP  AQGQFRIKDI  ERRKETSKSK  VKDPEKTNAS  KPTPEGNLFT
151   CKTPAIADIV  ILVDGSWSIG  RFNFRLVRLF  LENLVSAFNV  GSEKTRVGLA
201   QYSGDPRIEW  HLNAYGTKDA  VLDAVRNLPY  KGGNTLTGLA  LTYILENSFK
251   PEAGARPGVS  KIGILITDGK  SQDDVIPPAK  NLRDAGIELF  AIGVKNADIN
301   ELKEIASEPD  STHVYNVADF  NFMNSIVEGL  TRTVCSRVEE  QEKEIKGTIA
351   ASLGAPTDLV  TSDITARGFR  VSWTHSPGKV  EKYRVVYYPT  RGGQPEEVVV
401   DGSSSTAVLK  NLMSLTEYQI  AVFAIYSNAA  SEGLRGTETT  LALPMASDLK
451   LYDVSHSSMR  AKWNGVAGAT  GYMILYAPLT  EGLAADEKEI  KIGEASTELE
501   LDGLLPNTEY  TVTVYAMFGE  EASDPLTGQE  TTLPLSPPSN  LKFSDVGHNS
551   AKLTWDPASK  NVKGYRIMYV  KTDGTETNEV  EVGPVSTHTL  KSLTALTEYT
601   VAIFSLYDEG  QSEPLTGSFT  TRKVPPPQHL  EVDEASTDSF  RVSWKPTSSD
651   IAFYRLAWIP  LDGGESEEVV  LSGDADSYVI  EGLLPNTEYE  VSLLAVFDDE
701   TESEVVAVLG  ATIVGTTAIP  TTVTTTTTTT  ATTPKPTIAV  FRTGVRNLVI
751   DDETTSSLRV  VWDISDHNAQ  QFRVTYLTAK  GDRAEEAIMV  PGRQNTLLLQ
801   PLLPDTEYKV  TITPIYADGE  GVSVSAPGKT  LPLSAPRNLR  VSDEWYNRLR
851   ISWDAPPSPT  MGYRIVYKSI  NVPGPALETF  VGDDINTILI  LNLFSGTEYS
901   VKVFASYSTG  FSDALTGVAK  TLYLGVTNLD  TYQVRMTSLC  AQWQLHRHAT
951   AYRVVIESLV  DGKKQEVNLG  GGVPRHCFFE  LMPGTEYKIS  VHAQLQEIEG
1001  PAVSIMETTL  PFPTQPPTSP  STTLPPPTIP  PAKEVCKAAK  ADLVFLVDGS
1051  WSIGDDNFNK  IISFLYSTVG  ALDKIGPDGT  QVAIIQFSDD  PRTEFKLNAY
1101  KTKETLLEAI  QQIAYKGGNT  KTGKAIKHAR  EVLFTGEAGM  RKGIPKVLVV
1151  ITDGRSQDDV  NKVSREMQLD  GFSFFAIGVA  DADYSELVNI  GSKPSERHVF
1201  FVDDFDAFTK  IEDELITFVC  ETASATCPLV  FKDGDKLAGF  KMMEMFGLVE
1251  KEFSAIDGVS  MEPGTFNVYP  CYRLHKDALV  SQPTKYLHPE  GLPSDYTITF
1301  LFRILPDTPQ  EPFALWEILN  EQYEPLVGVI  LDNGGKTLTF  FNYDYKGDFQ
1351  TVTFEGPEIR  KIFYGSFHKL  HVVISKTTAK  IIIDCKEAGE  KTINAAGNIS
1401  SDGIEVLGRM  VRSRGPRDNS  APLQLQMFDI  VCATSWANRD  KCCELPGLRD
1451  EENCPALPHA  CSCSEANKGP  LGPPGPPGGP  GVRGAKGHRG  DPGPKGPDGP
1501  RGEIGVPGPQ  GPPGPQGPSG  LSIQGLPGPP  GEKGEKGDLG  FPGLQGVPGA
1551  SGSPGRDGAQ  GQRGLPGKDG  PTGPQGPPGP  VGIPGAPGVP  GITGSQGPQG
1601  DVGAPGAPGP  KGERGERGDL  QSQAMVRAVA  RQVCEQLIQG  HMARYNSILN
1651  QIPSQSVSTR  TIAGPPGEPG  RPGAPGPQGE  QGSPGMQGFP  GNPGQPGRPG
1701  ERGLPGEKGD  RGNPGVGTQG  PRGPPGSTGP  PGESRTGSPG  PPGSPGPRGP
1751  AGHTGPPGSQ  GPAGPPGYCD  PSSCAGYGMG  GGYGEPTDQD  IPVVQLPHNS
1801  YQIYDPEDLY  DGEQQPYVVH  GSYPLPSPYS  QSSYPSPHLA  QPEFTPVREE
1851  MEAVELRSPG  ISRFRRKIAK  RSIKTLEHKR  EMAKEPSQ
```

Structural and functional sites
Signal peptide: 1–28
NC3 domain: 29–1468
COL2 domain: 1469–1620
NC2 domain: 1621–1663
COL1 domain: 1664–1769
NC1 domain: 1770–1888
Alternatively spliced domain: 1813–1843 (residue 1812G becomes E)

Fibronectin type III repeats: 29–118, 352–441, 442–533, 534–622, 623–712, 741–831, 832–922, 923–1010
von Willebrand factor A repeats: 148–336, 1032–1221
PARP repeat: 1222–1468
Potential N-linked glycosylation sites: 138, 1398

Gene structure

The size and chromosomal localization of the α1(XIV) gene has not been determined.

References

[1] Dublet, B. and van der Rest, M. (1991) Type XIV collagen, a homotrimeric molecule extracted from fetal bovine skin and tendon, with a triple helical disulfide-bonded domain homologous to type IX and type XII collagens. J. Biol. Chem. 266: 6853–6858.

[2] Gordon, M.K. et al (1991) Cloning of a cDNA for a new member of the class of fibril-associated collagens with interrupted triple helices. Eur. J. Biochem. 201: 333–338.

[3] Just, M. et al (1991) Undulin is a novel member of the fibronectin-tenascin family of extracellular matrix glycoproteins. J. Biol. Chem. 266: 17326–17332.

[4] Aubert-Foucher, E. et al (1992) Purification and characterization of native type XIV collagen. J. Biol. Chem. 266: 15759–15764.

[5] Trueb, J. and Trueb, B. (1992) Type XIV collagen is a variant of undulin. Eur. J. Biochem. 207: 549–557.

[6] Watt, S.L. et al (1992) Characterization of collagen types XII and XIV from fetal bovine cartilage. J. Biol. Chem. 267: 20093–20099.

[7] Walchli, C. et al (1993) Complete primary structure of chicken collagen XIV. Eur. J. Biochem. 212: 483–490.

[8] Gerecke D.R. et al (1993) Type XIV collagen is encoded by alternative transcripts with distinct 5' regions and is a multidomain protein with homologies to von Willebrand's factor, fibronectin, and other matrix proteins. J. Biol. Chem. 268: 12177–12184.

Collagen type XV

Type XV collagen has been characterized at the cDNA and genomic level only. The molecule was identified from clones and transcripts of human placenta.

Molecular structure

The complete sequence of the α1(XV) chain has not been determined. The human placenta cDNA clone codes for a triple-helical domain of 557 residues and part of a non-collagenous COOH-terminal domain. The collagenous domain has 13 interruptions of 2–45 amino acids. Four of the interruptions contain potential sites for serine-linked glycosaminoglycans and asparagine-linked oligosaccharides. The molecular structure of type XV collagen is unknown [1-3].

Isolation

The type XV collagen protein has not been isolated. However, a collagenous polypeptide of approximately 125 kDa has been identified by Western blotting HeLa cell lysates with an antiserum raised to a synthetic peptide from the non-collagenous COOH-terminal domain.

Primary structure

Unknown.

Gene structure

The gene encoding the α1(XV) chain has been localized to chromosome 9 at locus q21–22. Partial analysis of the gene indicates that the sizes of the exons are highly variable (varying between 36 and 140 bp) [4].

References

[1] Kivirikko, S. et al (1992) Partial characterisation of the human gene for the α1 chain of the type XV collagen. Abstracts of the XIIIth Meeting of the Federation of European Connective Tissue Societies, 1/5.

[2] Myers, J. et al (1992) Identification of a previously unknown collagen chain, α1 (XV), characterised by extensive interruptions in the triple-helical region. Proc. Natl Acad. Sci. USA 89: 10144–10148.

[3] Pihlajaniemi, T. et al (1992) Partial characterisation of the α1 chain of type XV collagen and location of its transcripts in human tissues. Abstracts of the XIIIth Meeting of the Federation of European Connective Tissue Societies, 10/7.

[4] Huebner, K. et al (1992) Chromosomal assignment of a gene encoding a new collagen type (COL15A1) to 9q21–q22. Genomics 14: 220–224.

Collagen type XVI

Type XVI collagen has been characterized at the cDNA level only and appears to be a member of the FACIT collagens together with types IX, XII and XIV collagen. It is expressed in human skin and lung fibroblasts, keratinocytes, arterial smooth muscle cells and amnion.

Molecular structure

The predicted α1(XVI) chain comprises ten collagenous domains (COL1–10) that range in size from 15 to 422 amino acids and 11 relatively short (11–39 amino acids) non-collagenous domains (NC1–11). Thirty-two cysteine residues are present, mainly in the NC domains. The two cysteines present four amino acids apart at the COL1/NC1 junction and the size of the COL1 domain are characteristic of FACIT collagens. The molecular structure of type XVI collagen is unknown [1,2].

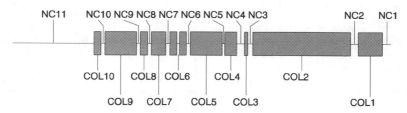

Isolation

The type XVI collagen protein has not been isolated. However, a collagenous polypeptide of approximately 160 kDa has been identified by Western blotting the culture medium of arterial smooth muscle cells using an antibody generated against a synthetic peptide from the NC3 domain.

Accession number

M92642

Primary structure: α1(XVI) chain

Ala	A	86	Cys	C	32	Asp	D	52	Glu	E	92

Ala A 86   Cys C 32   Asp D 52   Glu E 92
Phe F 30   Gly G 390   His H 19   Ile I 41
Lys K 88   Leu L 99   Met M 26   Asn N 23
Pro P 278   Gln Q 79   Arg R 63   Ser S 77
Thr T 44   Val V 63   Trp W 8   Tyr Y 13

Mol. wt (calc.) = 157 516   Residues = 1603

```
1    MWVSWAPGLW   LLGLWATFGH   GANTGAQCPP   SQQEGLKLEH   SSSLPANVTG
51   FNLIHRLSLM   KKSAIKKIRN   PKGPLILRLG   AAPVTQPTRR   VFPRGLPEEF
101  ALVLTLLLKK   HTHQKTWYLF   QVTDANGYPQ   ISLEVNSQER   SLELRAQGQD
151  GDFVSCIFPV   PQLFDLRWHK   LMLSVAGRVA   SVHVDCSSAS   SQPLGPRRPM
201  RPVGHVFLGL   DAEQGKPVSF   DLQQVHIYCD   PELVLEEGCC   EILPAGCPPE
251  TSKARRDTQS   NELIEINPQS   EGKVYTRCFC   LEEPQNSEVD   AQLTGRISQK
301  AERGAKVHQE   TAADECPPCV   HGARDSNVTL   APSGPKGGKG   ERGLPGPPGS
351  KGEKGARGND   CVRISPDAPL   QCAEGPKGEK   GESGALGPSG   LPGSTGEKGQ
401  KGEKGDGGIK   GVPGKPGRDA   PGEICVIGPK   GQKGDPGFVG   PEGLAGEPGP
```

83

```
451   PGLPGPPGIG   LPGTPGDPGG   PPGPKGDKGS   SGIPGKEGPG   GKPGKPGVKG
501   EKGDPCEVCP   TLPEGFQNFV   GLPGKPGPKG   EPGDPVRARG   DPGIQGIKGE
551   KGEPCLSCSS   VVGAQHLVSS   TGASGDVGSP   GFGLPGLPGR   AGVPGLKGEK
601   GNFGEAGPAG   SPGPPGPVGP   AGIKGAKGEP   CEPCPALSNL   QDGDVRVVAL
651   PGPSGEKGEP   GPPGFGLPGK   QGKAGERGLK   GQKGDAGNPG   DPGTPGTTGR
701   PGLSGEPGVQ   GPAGPKGEKG   DGCTACPSLQ   GTVTDMAGRP   GQPGPKGEQG
751   PEGVGRPGKP   GQPGLPGVQG   PPGLKGVQGE   PGPPGRGVQG   PQGEPGAPGL
801   PGIQGLPGPR   GPPGPTGEKG   AQGSPGVKGA   TGPVGPPGAS   VSGPPGRDGQ
851   QGQTGLRGTP   GEKGPRGEKG   EPGECSCPSQ   GDLIFSGMPG   APGLWMGSSW
901   QPGPQGPPGI   PGPPGPPGVP   GLQGVPGNNG   LPGQPGLTAE   LGSLPIEQHL
951   LKSICGDCVQ   GQRAHPGYLV   EKGEKGDQGI   PGVPGLDNCA   QCFLSLERPR
1001  AEEARGDNSE   GDPGCVGSPG   LPGPPGLPGQ   RGEEGPPGMR   GSPGPPGPIG
1051  PPGFPGAVGS   PGLPGLQGER   GLTGLTGDKG   EPGPPGQPGY   PGATGPPGLP
1101  GIKGERGYTG   SAGEKGEPGP   PGSEGLPGPP   GPAGPRGERG   PQGNSGEKGD
1151  QGFQGQPGFT   GPTGSPGFPG   KVGSPGPPGP   QAEKGSEGIR   GPSGLPGSPG
1201  PPGPPGIQGP   AGLDGLDGKD   GKPGLRGDPG   PAGPPGLMGP   PGFKGKTGHP
1251  GLPGPKGDCG   KPGPPGSTGR   PGAEGEPGAM   GPQGRPGPPG   HVGPPGPPGQ
1301  PGPAGISAVG   LKGDRGATGE   RGLAGLPGQP   GPPGHPGPPG   EPGTDGAAGK
1351  EGPPGKQGFY   GPPGPKGDPG   AAGQKGQAGE   KGRAGMPGGP   GKSGSMGPVG
1401  PPGPAGERGH   PGAPGPSGSP   GLPGVPGSMG   DMVNYDEIKR   FIRQEIIKMF
1451  DERMAYYTSR   MQFPMEMAAA   PGRPGPPGKD   GAPGRPGAPG   SPGLPGQIGR
1501  EGRQGLPGVR   GLPGTKGEKG   DIGIGIAGEN   GLPGPPGPQG   PPGYGKMGAT
1551  GPMGQQGIPG   IPGPPGPMGQ   PGKAGHCNPS   DCFGAMPMEQ   QYPPMKTMKG
1601  PFG
```

Structural and functional sites

Signal peptide: 1–21
PARP repeat 22–255
NC11 domain: 22–333
COL10 domain: 334–360
NC10 domain: 361–374
COL9 domain: 375–505
NC9 domain: 506–520
COL8 domain: 521–554
NC8 domain: 555–571
COL7 domain: 572–630
NC7 domain: 631–651
COL6 domain: 652–722
NC6 domain: 723–737
COL5 domain: 738–875
NC5 domain: 876–886
COL4 domain: 887–838
NC4 domain: 839–972
COL3 domain: 973–987
NC3 domain: 988–1010
COL2 domain: 1011–1432
NC2 domain: 1433–1471
COL1 domain: 1472–1577
NC1 domain: 1578–1603
Potential N-linked glycosylation sites: 47, 327, 1578

Gene structure

The gene for the α1(XVI) chain has been localized to human chromosome 1 at locus p34–35 [1].

References

[1] Pan, T.-C. et al (1992) Cloning and chromosomal location of human α1(XVI) collagen. Proc. Natl Acad. Sci. USA 89: 6565–6569.

[2] Yamaguchi, N. et al (1992) Molecular cloning and partial characterisation of a novel collagen chain, α1 (XVI), consisting of repetitive collagenous domains and cysteine-containing non-collagenous segments. J. Biochem. 112: 856–863.

Decorin is a member of the family of small CS/DS proteoglycans. Other members of this family include fibromodulin and biglycan. Decorin is relatively abundant in bone, tendon, sclera, skin, aorta and cornea. Rotary shadowing electron microscopy shows decorin containing one GAG chain. Decorin binds type I collagen and has been observed associated with collagen fibrils. Furthermore, decorin can inhibit collagen fibrillogenesis of both type I and II collagens *in vitro* in a manner that is not affected by removal of the GAG chain. Decorin can bind TGFβ and neutralize its biological activity. The decorin core protein binds with relatively high affinity ($K_d = 3 \times 10^{-7}$ M) to type VI collagen.

Molecular structure

Most of the decorin protein consists of 11 repeats of a 23-residue leucine-rich sequence. These repeats are homologous to sequences in other proteins including biglycan, fibromodulin, the serum protein LRG, platelet surface protein GPIb, ribonuclease/angiotensin inhibitor, chaoptin, toll protein and adenylate cyclase. Decorin, biglycan and fibromodulin are similar in size and their cysteine residues are located in conserved positions. There is preliminary evidence that decorin is synthesized as a procore protein *1–11*.

Isolation

Decorin can be isolated from a variety of tissues including articular cartilage by guanidine–HCl extraction, density gradient centrifugation, ion-exchange chromatography and octyl-Sepharose chromatography *12*.

Accession number

P07585

Primary structure

Sequence conflicts: 37 G to A
45 D to P

Ala	A	17	Cys	C	6	Asp	D	18	Glu	E	17

Ala A 17   Cys C 6   Asp D 18   Glu E 17
Phe F 13   Gly G 24   His H 9   Ile I 21
Lys K 27   Leu L 48   Met M 5   Asn N 27
Pro P 23   Gln Q 15   Arg R 13   Ser S 27
Thr T 16   Val V 23   Trp W 2   Tyr Y 8

Mol. wt (calc.) = 39 701   Residues = 359

```
1     MKATIILLLL   AQVSWAGPFQ   QRGLFDFMLE   DEASGIGPEV   PDDRDFEPSL
51    GPVCPFRCQC   HLRVVQCSDL   GLDKVPKDLP   PDTTLLDLQN   NKITEIKDGD
101   FKNLKNLHAL   ILVNNKISKV   SPGAFTPLVK   LERLYLSKNQ   LKELPEKMPK
151   TLQELRAHEN   EITKVRKVTF   NGLNQMIVIE   LGTNPLKSSG   IENGAFQGMK
201   KLSYIRIADT   NITSIPQGLP   PSLTELHLDG   NKISRVDAAS   LKGLNNLAKL
251   GLSFNSISAV   DNGSLANTPH   LRELHLDNNK   LTRVPGGLAE   HKYIQVVYLH
301   NNNISVVGSS   DFCPPGHNTK   KASYSGVSLF   SNPVQYWEIQ   PSTFRCVYVR
351   SAIQLGNYK
```

Structural and functional sites
Signal peptide: 1–16
Propeptide: 17–30
Leucine-rich repeats: 77–98, 99–122, 123–145, 146–167, 168–193, 194–217, 218–238, 239–262, 263–285, 286–308
Glycosaminoglycan attachment sites: 34
Potential N-linked glycosylation sites: 211, 262, 303

Gene structure

The human decorin gene is located on chromosome 12q21–q22 [11].

References
[1] Scott, J.E. and Orford, C.R. (1981) Dermatan sulfate-rich proteoglycan associates with rat tail tendon collagen at the d-band in the gap region. Biochem. J. 197: 213–216.

[2] Vogel, K.G. and Heinegård, D. (1985) Characterization of proteoglycans from adult bovine tendon. J. Biol. Chem. 260: 9298–9306.

[3] Krusius, T. and Ruoslahti, E. (1986) Primary structure of an extracellular matrix proteoglycan core protein deduced from cloned cDNA. Proc. Natl Acad. Sci. USA 83: 7683–7687.

[4] Scott, P.G. et al (1986) A role for disulphide bridges in the protein core in the interaction of proteodermatan sulfate and collagen. Biochem. Biophys. Res. Commun. 138: 1348–1354.

[5] Roughley, P.J. and White, R.J. (1989) Dermatan sulphate proteoglycans of human articular cartilage. The properties of dermatan sulphate proteoglycans I and II. Biochem. J. 262: 823–827.

[6] Vogel, K.G. and Brown, D.C. (1990) Characteristics of the in vitro interaction of a small proteoglycan (PG-II) of bovine tendon with type I collagen. Matrix 9: 468–478.

[7] Yamaguchi, Y.et al (1990) Negative regulation of transforming growth factor β by the proteoglycan decorin. Nature 346: 281–284.

[8] Sawhney, R.S. et al (1991) Biosynthesis of small proteoglycan II (decorin) by chondrocytes and evidence for a procore protein. J. Biol. Chem. 266: 9231–9240.

[9] Bidanset, D.J. et al (1992) Binding of the proteoglycan decorin to collagen type VI. J. Biol. Chem. 267: 5250–5256.

[10] Border, W.A. et al (1992) Natural inhibition of transforming growth factor β protects against scarring in experimental kidney disease. Nature 360: 361–364.

[11] Pulkkinen, L. et al (1992) Expression of decorin in human tissues and cell lines and defined chromosomal assignment of the gene locus (DCN). Cytogenet. Cell Genet., 60: 107–111.

[12] Choi, H.U. et al (1989) Characterization of the dermatan sulfate proteoglycans, DS-PG1 and DS-PGII, from bovine articular cartilage and skin isolated by octyl-Sepharose chromatography. J. Biol. Chem. 264: 2876–2884.

Elastin

Elastin is the major protein of the elastic fibres that form a randomly oriented, interconnected network in many tissues. Elastin content may vary from 2% of dry weight in skin to over 70% in the nuchal ligament of grazing animals. A high content of hydrophobic amino acids makes elastin one of the most chemically and proteinase-resistant proteins in the body. Its principal function is to provide elasticity and resilience to tissues; however, elastin promotes cell adhesion and elastin peptides have been shown to be chemotactic.

Molecular structure

Tropoelastin, the soluble precursor of elastin, is a single polypeptide chain; however, alternative splicing produces a number of different isoforms. Splicing is regulated in a developmental and/or tissue-specific manner. Individual chains are secreted into the extracellular space and, in association with microfibrillar components, assemble to form elastic fibres. The elastin polypeptide is made up of a number of alternating hydrophobic and cross-link repeats. Deamination of specific lysines by lysyl oxidase allows the introduction of covalent cross-links to stabilize the elastic fibres. These cross-links (desmosine and isodesmosine) are specific to elastic fibres. Lysines in cross-link repeats usually occur in pairs, but in two instances three lysines are found near one another (residue numbers 375–382 and 558–567). Strong homology exists between the human, bovine and porcine sequences, but these differ considerably from the chicken. Several VGVAPG sequences and their homologues are reported to mediate cell adhesion via a 67-kDa binding protein [1-5].

Isolation

Early methods for preparing elastin from tissues relied on its chemical resistance to reflux in 0.1 M NaOH. The absence of methionine also allows CNBr digestion of the tissue under denaturing conditions to remove other proteins [6].

Accession number

A32707; P15502

Primary structure

Ala	A	167	Cys	C	2	Asp	D	3	
Phe	F	16	Gly	G	223	His	H	2	
Lys	K	35	Leu	L	52	Met	M	1	
Pro	P	100	Gln	Q	10	Arg	R	12	
Thr	T	12	Val	V	98	Trp	W	0	

Glu	E	5
Ile	I	17
Asn	N	0
Ser	S	16
Tyr	Y	15

Mol. wt (calc.) = 68 419   Residues = 786

```
1    MAGLTAAAPR   PGVLLLLLSI   LHPSRPGGVP   GAIPGGVPGG   VFYPGAGLGA
51   LGGGALGPGG   KPLKPVPGGL   AGAGLGAGLG   AFPAVTFPGA   LVPGGVADAA
101  AAYKAAKAGA   GLGGVPGVGG   LGVSAGAVVP   QPGAGVKPGK   VPGVGLPGVY
```

151	PGGVLPGARF	PGVGVLPGVP	TGAGVKPKAP	GVGGAFAGIP	GVGPFGGPQP
201	GVPLGYPIKA	PKLPGGYGLP	YTTGKLPYGY	GPGGVAGAAG	KAGYPTGTGV
251	GPQAAAAAAA	KAAAKFGAGA	AGVLPGVGGA	GVPGVPGAIP	GIGGIAGVGT
301	PAAAAAAAAA	AKAAKYGAAA	GLVPGGPGFG	PGVVGVPGAG	VPGVGVPGAG
351	IPVVPGAGIP	GAAVPGVVSP	EAAAKAAAKA	AKYGARPGVG	VGGIPTYGVG
401	AGGFPGFGVG	VGGIPGVAGV	PSVGGVPGVG	GVPGVGISPE	AQAAAAAKAA
451	KYGAAGAGVL	GGLVPGPQAA	VPGVPGTGGV	PGVGTPAAAA	AKAAAKAAQF
501	GLVPGVGVAP	GVGVAPGVGV	APGVGLAPGV	GVAPGVGVAP	GVGVAPGIGP
551	GGVAAAAKSA	AKVAAKAQLR	AAAGLGAGIP	GLGVGVGVPG	LGVGAGVPGL
601	GVGAGVPGFG	AGADEGVRRS	LSPELREGDP	SSSQHLPSTP	SSPRVPGALA
651	AAKAAKYGAA	VPGVLGGLGA	LGGVGIPGGV	VGAGPAAAAA	AAKAAAKAAQ
701	FGLVGAAGLG	GLGVGGLGVP	GVGGLGGIPP	AAAAKAAKYG	AAGLGGVLGG
751	AGQFPLGGVA	ARPGFGLSPI	FPGGACLGKA	CGRKRK	

Structural and functional sites

Signal peptide: 1–26

Hydrophobic repeats: 28–53, 66–77, 109–124, 143–158, 181–190, 215–228, 229–248, 267–296, 317–365, 384–438, 453–481, 501–548, 570–644, 658–681, 702–726, 740–757, 758–772

Cross-link repeats: 54–65, 78–108, 125–142, 159–180, 191–214, 249–266, 297–316, 366–383, 439–452, 482–500, 549–569, 645–657, 682–701, 727–739

Alternatively spliced repeats: 453–481, 482–500, 501–548, 570–644, 740–757, 758–772

Free cysteine residues: 781, 776

β spiral motif: 505–546

Gene structure

Human elastin is encoded by a single-copy gene found on chromosome 7 (at locus q11.2). The gene is approximately 45 kb long and contains 34 exons separated by large introns. The hydrophobic and cross-link repeats of the protein are encoded by distinct exons. In most cases, alternative splicing either includes or deletes an exon in a cassette-like fashion. In two cases, splicing occurs within an exon (501–548 and 612–644). At the exon–intron border, codons are split in the same way; the 5′ border provides the second and third nucleotides while the first nucleotide is found at the 3′ border [1,3,4,5,7].

References

[1] Sandberg, L.B. and Davidson, J.M. (1984) Elastin and its gene. Peptide Protein Rev. 3: 169–193.

[2] Senior, R.M. et al (1984) Val-Gly-Val-Ala-Pro-Gly, a repeating peptide in elastin, is chemotactic for fibroblasts and monocytes. J. Cell Biol. 99: 870–874.

[3] Yeh, H. et al (1989) Structure of the bovine elastin gene and S1 nuclease analysis of the alternative splicing of elastin mRNA in the bovine nuchal ligament. Biochemistry 28: 2365–2370.

[4] Indik, Z. et al (1990) Structure of the elastin gene and alternative splicing of elastin mRNA. In: Extracellular Matrix Genes. Sandell, L.J. and Boyd C.D., eds, Academic Press, New York, pp. 221–250.

5 Pollock, J et al (1990) Chick tropoelastin isoforms. From the gene to the extracellular matrix. J. Biol. Chem. 265: 3697–3702.

6 Soskel, N.T. et al (1987) Isolation and characterisation of insoluble and soluble elastins. Methods Enzymol. 144: 196–214.

7 Fazio, M.J et al (1991) Human elastin gene: New evidence for localization to the long arm of chromosome 7. Am. J. Hum. Genet. 48: 696–703.

Entactin   nidogen

Entactin/nidogen is a sulphated glycoprotein that is an integral component of basement membranes. It associates specifically with both laminin (in a 1:1 molar ratio) and type IV collagen and is thought to play an important role in linking these two molecules together in the basement membrane.

Molecular structure

Entactin/nidogen is a single-chain polypeptide with a molecular weight of approximately 150 000. Isolated under denaturing conditions, the molecule resembles a dumb-bell when viewed by electron microscopy after rotary shadowing. However, recombinant molecules isolated from tissue culture medium under non-denaturing conditions appear to contain three rather than two globular domains separated by rod-like segments. The entactin/nidogen molecule contains a series of six-cysteine EGF repeats, broken up by repeats found in thyroglobulin and the LDL receptor [1,2].

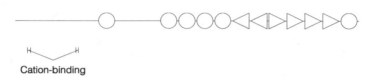

Cation-binding

Isolation

Entactin/nidogen can be isolated from basement membranes in association with laminin by extraction with chelating agents. The entactin/nidogen–laminin complex can then be dissociated with buffers containing guanidine–HCl [3].

Accession number

P14543

Primary structure

Ala	A	79	Cys	C	50	Asp	D	76	Glu	E	73
Phe	F	53	Gly	G	113	His	H	34	Ile	I	56
Lys	K	29	Leu	L	81	Met	M	8	Asn	N	45
Pro	P	87	Gln	Q	58	Arg	R	73	Ser	S	93
Thr	T	96	Val	V	86	Trp	W	13	Tyr	Y	44

Mol. wt (calc.) = 136 337   Residues = 1247

```
1     MLASSSRIRA  AWTRALLLPL  LLAGPVGCLS  RQELFPFGPG  QGDLELEDGD
51    DFVSPALELS  GALRFYDRSD  IDAVYVTTNG  IIATSEPPAK  ESHPGLFPPT
101   FGAVAPFLAD  LDTTDGLGKV  YYREDLSPSI  TQRAAECVHR  GFPEISFQPS
151   SAVVVTWESV  APYQGPSRDP  DQKGKRNTFQ  AVLASSDSSS  YAIFLYPEDG
201   LQFHTTFSKK  ENNQVPAVVA  FSQGSVGFLW  KSNGAYNIFA  NDRESIENLA
251   KSSNSGQQGV  WVFEIGSPAT  TNGVVPADVI  LGTEDGAEYD  DEDEDYDLAT
301   TRLGLEDVGT  TPFSYKALRR  GGADTYSVPS  VLSPRRAATE  RPLGPPTERT
351   RSFQLAVETF  HQQHPQVIDV  DEVEETGVVF  SYNTDSRQTC  ANNRHQCSVH
```

```
401    AECRDYATGF    CCSCVAGYTG    NGRQCVAEGS    PQRVNGKVKG    RIFVGSSQVP
451    IVFENTDLHS    YVVMNHGRSY    TAISTIPETV    GYSLLPLAPV    GGIIGWMFAV
501    EQDGFKNGFS    ITGGEFTRQA    EVTFVGHPGN    LVIKQRFSGI    DEHGHLTIDT
551    ELEGRVPQIP    FGSSVHIEPY    TELYHYSTSV    ITSSSTREYT    VTEPERDGAS
601    PSRIYTYQWR    QTITFQECVH    DDSRPALPST    QQLSVDSVFV    LYNQEEKILR
651    YAFSNSIGPV    REGSPDALQN    PCYIGTHGCD    TNAACRPGPR    TQFTCECSIG
701    FRGDGRTCYD    IDECSEQPSV    CGSHTICNNH    PGTFRCECVE    GYQFSDEGTC
751    VAVVDQRPIN    YCETGLHNCD    IPQRAQCIYT    GGSSYTCSCL    PGFSGDGQAC
801    QDVDECQPSR    CHPDAFCYNT    PGSFTCQCKP    GYQGDGFRCV    PGEVEKTRCQ
851    HEREHILGAA    GATDPQRPIP    PGLFVPECDA    HGHYAPTQCH    GSTGYCWCVD
901    RDGREVEGTR    TRPGMTPPCL    STVAPPIHQG    PAVPTAVIPL    PPGTHLLFAQ
951    TGKIERLPLE    GNTMRKTEAK    AFLHVPAKVI    IGLAFDCVDK    MVYWTDITEP
1001   SIGRASLHGG    EPTTIIRQDL    GSPEGIAVDH    LGRNIFWTDS    NLDRIEVAKL
1051   DGTQRRVLFE    TDLVNPRGIV    TDSVRGNLYW    TDWNRDNPKI    ETSYMDGTNR
1101   RILVQDDLGL    PNGLHFDAFS    SQLCWVDAGT    NRAECLNPSQ    PSRRKALEGL
1151   QYPFAVTSYG    KNLYFTDWKM    NSVVALDLAI    SKETDAFQPH    KQTRLYGITT
1201   ALSQCPQGHN    YCSVNNGGCT    HLCLATPGSR    TCRCPDNTLG    VDCIERK
```

Structural and functional sites

Signal peptide: 1–28
EGF (6C) repeats: 388–425, 670–711, 712–759, 760–803, 804–846, 1212–1247
Thyroglobulin repeats: 847–884, 885–921
LDL receptor repeats: 983–1028, 1029–1078, 1079–1128, 1129–1174
EF-hand-type divalent cation binding sites: 43–54, 278–289
Potential O-linked sulphation sites: 289, 296
Potential N-linked glycosylation site: 1137
Transglutaminase cross-linking site (to laminin): 756 (equivalent to Q726 in murine sequence which is a proven cross-linking site)
RGD cell adhesion site: 702–704 (not conserved in murine sequence)

Gene structure

The entactin/nidogen gene has been localized to human chomosome 1q43 [2].

References

[1] Nagayoshi, T. et al (1989) Human nidogen: Complete amino acid sequence and structural domains deduced from cDNAs, and evidence for polymorphism of the gene. DNA 8: 581–594.
[2] Olsen, D.R. et al (1989) Human nidogen: cDNA cloning, cellular expression, and mapping of the gene to chromosome 1q43. Am. J. Hum. Genet. 44: 876–855.
[3] Paulsson, M. et al (1987) Laminin-nidogen complex. Extraction with chelating agents and structural characterisation. Eur. J. Biochem. 166: 467–478.

Fibrillin

The fibrillins are glycoproteins which occur as a major component of a subset of connective tissue microfibrils. These microfibrils have a beaded appearance and a cross-sectional diameter of 10–12 nm. In elastic tissues, these structures are thought to provide the scaffold onto which elastin is assembled to form elastic fibres, although their functional role in non-elastic tissues is unclear. Defects in fibrillin on human chromosome 15 have been shown to result in the Marfan syndrome.

Molecular structure

The fibrillin transcribed from chromosome 15 has a molecular weight of approximately 312 000 and contains 2871 amino acids. Its high cysteine content (14%) suggests that intrachain disulphide bonds are important in stabilizing the molecule. Five structurally distinct regions have been delineated, the largest of which contains a series of cysteine-rich repeats. Internal sequence homology reveals two distinct motifs that make up the majority of the protein; there are 41 six-cysteine EGF repeats and 8 eight-cysteine repeats related to motifs found in the TGFβ1-binding protein. There is a second cysteine-rich region towards the NH_2-terminus which, in addition to two EGF repeats, contains a number of unusual cysteine-rich modules that may be unique to this molecule. Separating these regions is a polypeptide domain rich in proline (42%). In contrast, the COOH-terminal 184 amino acids, which are rich in lysine and arginine, have no homology to the rest of the protein, and the extreme NH_2-terminal 29 amino acids are highly basic. The length of the molecule has been reported to be 148 nm, but it is unclear how this soluble precursor molecule assembles into microfibrils with a periodicity of 50 nm. Sedimentation experiments show that the molecule is extremely stable even in 8 M urea. One report suggests that the fibrillin molecule is processed and deposited in the cell-associated matrix as a smaller protein (approximate molecular weight of 320 000). A putative RGD cell adhesion site is located in one of the eight-cysteine repeats [1-7].

Isolation

Fibrillin has been purified from culture medium by precipitation with 40% ammonium sulphate, DEAE ion-exchange chromatography, and FPLC on Mono-Q and Superose 6 columns [8]. Fibrillin-containing microfibrils have been isolated by bacterial collagenase digestion and Sepharose CL-2B chromatography [9].

Accession number

S17064

Primary structure: fibrillin 1 (FBN1)

Ala	A	89	Cys	C	362	Asp	D	172	Glu	E	201
Phe	F	84	Gly	G	307	His	H	48	Ile	I	148
Lys	K	111	Leu	L	141	Met	M	52	Asn	N	188
Pro	P	175	Gln	Q	101	Arg	R	131	Ser	S	173
Thr	T	166	Val	V	108	Trp	W	13	Tyr	Y	94

Mol. wt (calc.) = 311 949   Residues = 2871

```
1      MRRGRLLEIA   LGFTVLLASY   TSHGADANLE   AGNVKETRAS   RAKRRGGGGH
51     DALKGPNVCG   SRYNAYCCPG   WKTLPGGNQC   IVPICRHSCG   DGFCSRPNMC
101    TCPSGQIAPS   CGSRSIQHCN   IRCMNGGSCS   DDHCLCQKGY   IGTHCGQPVC
151    ESGCLNGGRC   VAPNRCACTY   GFTGPQCERD   YRTGPCFTVI   SNQMCQGQLS
201    GIVCTKQLCC   ATVGRAWGHP   CEMCPAQPHP   CRRGFIPNIR   TGACQDVDEC
251    QAIPGLCQGG   NCINTVGSFE   CKCPAGHKLN   EVSQKCEDID   ECSTIPGICE
301    GGECTNTVSS   YFCKCPPGFY   TSPDGTRCID   VRPGYCYTAL   TNGRCSNQLP
351    QSITKMQCCC   DAGRCWSPGV   TVAPEMCPIR   ATEDFNKLCS   VPMVIPGRPE
401    YPPPPLGPIP   PVLPVPPGFP   PGPQIPVPRP   PVEYLYPSRE   PPRVLPVNVT
451    DYCQLVRYLC   QNGRCIPTPG   SYRCECNKGF   QLDLRGECID   VDECEKNPCA
501    GGECINNQGS   YTCQCRAGYQ   STLTRTECRD   IDECLQNGRI   CNNGRCINTD
551    GSFHCVCNAG   FHVTRDGKNC   EDMDECSIRN   MCLNGMCINE   DGSFKCICKP
601    GFQLASDGRY   CKDINECETP   GICMNGRCVN   TDGSYRCECF   PGLAVGLDGR
651    VCVDTHMRST   CYGGYKRGQC   IKPLFGAVTK   SECCCASTEY   AFGEPCQPCP
701    AQNSAEYQAL   CSSGPGMTSA   GSDINECALD   PDICPNGICE   NLRGTYKCIC
751    NSGYEVDSTG   KNCVDINECV   LNSLLCDNGQ   CRNTPGSFVC   TCPKGFIYKP
801    DLKTCEDIDE   CESSPCINGV   CKNSPGSFIC   ECSSESTLDP   TKTICIETIK
851    GTCWQTVICG   RCEININGAT   LKSQCCSSLG   AAWGSPCTLC   QVDPICGKGY
901    SRIKGTQCED   IDECEVFPGV   CKNGLCVNTR   GSFKCQCPSG   MTLDATGRIC
951    LDIRLETCFL   RYEDEECTLP   IAGRHRMDAC   CCSVGAAWGT   EECEECPMRN
1001   TPEYEELCPR   GPGFATKEIT   NGKPFFKDIN   ECKMIPSLCT   HGKCRNTIGS
1051   FKCRCDSGFA   LDSEERNCTD   IDECRISPDL   CGRGQCVNTP   GDFECKCDEG
1101   YESGFMMMKN   CMDIDECQRD   PLLCRGGVCH   NTEGSYRCEC   PPGHQLSPNI
1151   SACIDINECE   LSAHLCPNGR   CVNLIGKYQC   ACNPGYHSTP   DRLFCVDIDE
1201   CSIMNGGCET   FCTNSEGSYE   CSCQPGFALM   PDQRSCTDID   ECEDNPNICD
1251   GGQCTNIPGE   YRCLCYDGFM   ASEDMKTCVD   VNECDLNPNI   CLSGTCENTK
1301   GSFICHCDMG   YSGKKGKTGC   TDINECEIGA   HNCGKHAVCT   NTAGSFKCSC
1351   SPGWIGDGIK   CTDLDECSNG   THMCSQHADC   KNTMGSYRCL   CKEGYTGDGF
1401   TCTDLDECSE   NLNLCGNGQC   LNAPGGYRCE   CDMGFVPSAD   GKACEDIDEC
1451   SLPNICVFGT   CHNLPGLFRC   ECEIGYELDR   SGGNCTDVNE   CLDPTTCISG
1501   NCVNTPGSYI   CDCPPDFELN   PTRVGCVDTR   SGNCYLDIRP   RGDNGDTACS
1551   NEIGVGVSKA   SCCCSLGKAW   GTPCEMCPAV   NTSEYKILCP   GGEGFRPNPI
1601   TVILEDIDEC   QELPGLCQGG   KCINTFGSFQ   CRCPTGYYLN   EDTRVCDDVN
1651   ECETPGICGP   GTCYNTVGNY   TCICPPDYMQ   VNGGNNCMDM   RRSLCYRNYY
1701   ADNQTCDGEL   LFNMTKKMCC   CSYNIGRAWN   KPCEQCPIPS   TDEFATLCGS
1751   QRPGFVIDIY   TGLPVDIDEC   REIPGVCENG   VCINMVGSFR   CECPVGFFYN
1801   DKLLVCEDID   ECQNGPVCQR   NAECINTAGS   YRCDCKPGYR   FTSTGQCNDR
1851   NECQEIPNIC   SHGQCIDTVG   SFYCLCHTGF   KTNDDQTMCL   DINECERDAC
1901   GNGTCRNTIG   SFNCRCNHGF   ILSHNNDCID   VDECASGNGN   LCRNGQCINT
1951   VGSFQCQCNE   GYEVAPDGRT   CVDINECLLE   PRKCAPGTCQ   NLDGSYRCIC
2001   PPGYSLQNEK   CEDIDECVEE   PEICALGTCS   NTEGSFKCLC   PEGFSLSSSG
2051   RRCQDLRMSY   CYAKFEGGKC   SSPKSRNHSK   QECCCALKGE   GWGDPCELCP
```

```
2101  TEPDEAFRQI  CPYGSGIIVG  PDDSAVDMDE  CKEPDVCKHG  QCINTDGSYR
2151  CECPFGYTLA  GNECVDTDEC  SVGNPCGNGT  CKNVIGGFEC  TCEEGFEPGP
2201  MMTCEDINEC  AQNPLLCAFR  CVNTYGSYEC  KCPVGYVLRE  DRRMCKDEDE
2251  CEEGKHDCTE  KQMECKNLIG  TYMCICGPGY  QRRPDGEGCV  DENECQTKPG
2301  ICENGRCLNT  RGSYTCECND  GFTASPNQDE  CLDNREGYCF  TEVLQNMCQI
2351  GSSNRNPVTK  SECCCDGGRG  WGPHCEICPF  QGTVAFKKLC  PHGRGFMTNG
2401  ADIDECKVIH  DVCRNGECVN  DRGSYHCICK  TGYTPDITGT  SCVDLNECNQ
2451  APKPCNFICK  NTEGSYQCSC  PKGYILQEDG  RSCKDLDECA  TKQHNCQFLC
2501  VNTIGGFTCK  CPPGFTQHHT  SCIDNNECTS  DINLCGSKGI  CQNTPGSFTC
2551  ECQRGFSLDQ  TGSSCEDVDE  CEGNHRCQHG  CQNIIGGYRC  SCPQGYLQHY
2601  QWNQCVDENE  CLSAHICGGA  SCHNTLGSYK  CMCPAGFQYE  QFSGGCQDIN
2651  ECGSAQAPCS  YGCSNTEGGY  LCGCPPGYFR  IGQGHCVSGM  GMGRGNPEPP
2701  VSGEMDDNSL  SPEACYECKI  NGYPKRGRKR  RSTNETDASN  IEDQSETEAN
2751  VSLASWDVEK  TAIFAFNISH  VSNKVRILEL  LPALTTLTNH  NRYLIESGNE
2801  DGFFKINQKE  GISYLHFTKK  KPVAGTYSLQ  ISSTPLYKKK  ELNQLEDKYD
2851  KDYLSGELGD  NLKMKIQVLL  H
```

Structural and functional sites

Signal peptide: 1–27

EGF (6C) repeats: 245–287, 288–329, 489–528, 529–571, 572–612, 613–653, 723–764, 765–806, 807–846, 910–951, 1028–1069, 1070–1112, 1113–1154, 1155–1196, 1197–1237, 1238–1279, 1280–1321, 1322–1362, 1363–1403, 1404–1445, 1446–1486, 1487–1527, 1606–1647, 1648–1688, 1766–1807, 1808–1848, 1849–1890, 1891–1929, 1930–1972, 1973–2012, 2013–2054, 2127–2165, 2166–2205, 2206–2246, 2247–2290, 2291–2362, 2402–2443, 2444–2484, 2485–2523, 2524–2566, 2567–2607

TGFβ1 receptor repeats: 654–722, 952–1027, 1528–1606, 1689–1765, 2055–2126, 2363–2401

Potential N-linked glycosylation sites: 448, 1067, 1149, 1369, 1484, 1581, 1669, 1703, 1902, 2077, 2178, 2734, 2750, 2767

RGD cell adhesion site: 1541–1543

Gene structure

Genes for fibrillin are located on human chromosomes 15 (FBN1) and 5 (FBN2). Intron and exon sizes have not been delineated, but the fibrillin 15 gene contains 65 exons. Fibrillin cDNA clones hybridize to a single mRNA of about 10 kb [2,4,5,7].

References

[1] Sakai, L.Y. et al (1986) Fibrillin, a new 350-kD glycoprotein, is a component of extracellular microfibrils. J. Cell Biol. 103: 2499–2509.

[2] Kainulainen, K. et al (1990) Location on chromosome 15 of the gene defect causing Marfan syndrome. N. Engl. J. Med. 323: 935–939.

[3] Dietz, H.C. et al (1991) The Marfan syndrome locus: confirmation of assignment to chromosome 15 and identification of tightly linked markers at 15q15–q21.3. Genomics 9: 355–361.

[4] Lee, B. et al (1991) Linkage of Marfan syndrome and a phenotypically related disorder to two different fibrillin genes. Nature 352: 330–334.

5 Maslen, C.L. et al (1991) Partial sequence of a candidate gene for the Marfan syndrome. Nature 352: 334–337.

6 Milewicz, D.M. et al (1992) Marfan syndrome: Defective synthesis, secretion and extracellular matrix formation of fibrillin by cultured dermal fibroblasts. J. Clin. Invest. 89: 79–86.

7 Pereira, L. et al (1993) Genomic organization of the sequence coding for fibrillin, the defective gene product in Marfan syndrome. Mol. Genet. (in press).

8 Sakai, L.Y. et al (1991) Purification and partial characterisation of fibrillin, a cysteine-rich structural component of connective tissue microfibrils. J. Biol. Chem. 266: 14763–14770.

9 Kielty, C.M. et al (1991) Isolation and ultrastructural analysis of microfibrillar structures from foetal bovine elastic tissues. J. Cell Sci. 99: 797–807.

Fibrinogen

Fibrinogen is a soluble plasma protein that is cleaved by thrombin to produce an insoluble fibrin clot. Polymerization to form fibrin acts also as a co-factor in platelet aggregation. Thrombin action cleaves fibrinopeptides A and B from the α and β chains, respectively, and exposes the polymerization site. The initial clot formed by polymerization is converted into a reinforced clot by factor XIIIa transglutaminase which catalyses the ϵ-(γ-glutamyl)lysine cross-linking between the γ chains and α chains of different monomers.

Molecular structure

Fibrinogen is a hexamer that contains two sets of non-identical chains (α, β and γ) linked to each other in an anti-parallel arrangement by disulphide bonds. All NH$_2$-termini of the chains are contained in a central nodule, two three-chain coiled-coils extend from this central nodule to two terminal domains and are gathered together at two nodes by disulphide rings. A further set of disulphide rings terminates the coiled-coil domain. The two terminal domains are composed of the COOH-two-thirds of the β and γ chains and are highly folded. A short section of the α chain is associated with these terminal domains before emerging into a highly flexible appendage that is readily cleaved by proteases. The α chain protruberance and the coiled-coils together account for half the mass of the fibrinogen molecule. Altogether, there are four carbohydrate clusters, one each on the two β and two γ chains. The α chain can be cross-linked in several places to fibronectin. Although the molecule contains two RGD sequences in its α chain, a pentapeptide QAGDV at the extreme COOH-terminus of the γ chain represents the main binding site for the platelet integrin receptor IIbIIIa (αIIbβ3). Interestingly, the γ chain can be alternatively spliced to produce a variant (γB) which lacks the COOH-terminal four residues of γA but adds 20 different residues unrelated to the integrin-binding motif. The γB chain is present in about 10% of fibrinogen molecules in plasma, but is absent from platelet fibrinogen. Variations in sequence at the fibrinopeptide A cleavage site (Arg35) lead to α-dysfibrinogenemias [1-6].

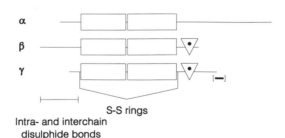

α
β
γ

S-S rings

Intra- and interchain
disulphide bonds

Isolation

Fibrinogen can be purified by a modified cold ethanol fractionation [7], and the individual chains separated by carboxymethylation in 6 M guanidine–HCl and CM-cellulose chromatography in 8 M urea [8].

Accession number
P02671

Primary structure: α chain

Sequence conflicts: 215–216 SR to RS
                 299 S to G
                 304 S to G
                 317–318 GT to SG
                 331 T to A

Ala	A	23	Cys	C	9	Asp	D	35	Glu	E	44

Ala A 23 Cys C 9 Asp D 35 Glu E 44
Phe F 20 Gly G 72 His H 16 Ile I 19
Lys K 40 Leu L 33 Met M 12 Asn N 29
Pro P 38 Gln Q 18 Arg R 42 Ser S 91
Thr T 51 Val V 32 Trp W 11 Tyr Y 9
Mol. wt (calc.) = 69 680   Residues = 644

```
1    MFSMRIVCLV  LSVVGTAWTA  DSGEGDFLAE  GGGVRGPRVV  ERHQSACKDS
51   DWPFCSDEDW  NYKCPSGCRM  KGLIDEVNQD  FTNRINKLKN  SLFEYQKNNK
101  DSHSLTTNIM  EILRGDFSSA  NNRDNTYNRV  SEDLRSRIEV  LKRKVIEKVQ
151  HIQLLQKNVR  AQLVDMKRLE  VDIDIKIRSC  RGSCSRALAR  EVDLKDYEDQ
201  QKQLEQVIAK  DLLPSRDRQH  LPLIKMKPVP  DLVPGNFKSQ  LQKVPPEWKA
251  LTDMPQMRME  LERPGGNEIT  RGGSTSYGTG  SETESPRNPS  SAGSWNSGSS
301  GPGSTGNRNP  GSSGTGGTAT  WKPGSSGPGS  TGSWNSGSSG  TGSTGNQNPG
351  SPRPGSTGTW  NPGSSERGSA  GHWTSESSVS  GSTGQWHSES  GSFRPDSPGS
401  GNARPNNPDW  GTFEEVSGNV  SPGTRREYHT  EKLVTSKGDK  ELRTGKEKVT
451  SGSTTTTRRS  CSKTVTKTVI  GPDGHKEVTK  EVVTSEDGSD  CPEAMDLGTL
501  SGIGTLDGFR  HRHPDEAAFF  DTASTGKTFP  GFFSPMLGEF  VSETESRGSE
551  SGIFTNTKES  SSHHPGIAEF  PSRGKSSSYS  KQFTSSTSYN  RGDSTFESKS
601  YKMADEAGSE  ADHEGTHSTK  RGHAKSRPVR  GIHTSPLGKP  SLSP
```

Structural and functional sites
Signal peptide: 1–19
Fibrinopeptide: 20–35
Phosphorylation site: 22
Thrombin cleavage site: 35–36
Polymerization site: 36–38
RGD cell adhesion sites: 114–116, 591–593
α2-plasmin inhibitor-binding site: 322
Acceptor cross-linking sites: 347, 385

Accession number
P02675

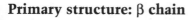
Primary structure: β chain

Sequence conflicts: 138–139 SQ to QS
145–146 FQ to QF
192 A to P

Ala	A	24	Cys	C	12	Asp	D	28	Glu	E	30
Phe	F	12	Gly	G	42	His	H	9	Ile	I	16
Lys	K	36	Leu	L	36	Met	M	18	Asn	N	32
Pro	P	22	Gln	Q	26	Arg	R	28	Ser	S	34
Thr	T	23	Val	V	28	Trp	W	14	Tyr	Y	21

Mol. wt (calc.) = 55 839   Residues = 491

```
1     MKRMVSWSFH   KLKTMKHLLL   LLLCVFLVKS   QGVNDNEEGF   FSARGHRPLD
51    KKREEAPSLR   PAPPPISGGG   YRARPAKAAA   TQKKVERKAP   DAGGCLHADP
101   DLGVLCPTGC   QLQEALLQQE   RPIRNSVDEL   NNNVEAVSQT   SSSSFQYMYL
151   LKDLWQKRQK   QVKDNENVVN   EYSSELEKHQ   LYIDETVNSN   IATNLRVLRS
201   ILENLRSKIQ   KLESDVSAQM   EYCRTPCTVS   CNIPVVSGKE   CEEIIRKGGE
251   TSEMYLIQPD   SSVKPYRVYC   DMNTENGGWT   VIQNRQDGSV   DFGRKWDPYK
301   QGFGNVATNT   DGKNYCGLPG   EYWLGNDKIS   QLTRMGPTEL   LIEMEDWKGD
351   KVKAHYGGFT   VQNEANKYQI   SVNKYRGTAG   NALMDGASQL   MGENRTMTIH
401   NGMFFSTYDR   DNDGWLTSDP   RKQCSKEDGG   GWWYNRCHAA   NPNGRYYWGG
451   QYTWDMAKHG   TDDGVVWMNW   KGSWYSMRKM   SMKIRPFFPQ   Q
```

Structural and functional sites

Signal peptide: 1–30
Fibrinopeptide: 31–44
Pyrrolidone carboxylic acid: 31
Potential N-linked glycosylation site: 394
Thrombin cleavage site: 44–45
Possible polymerization site: 45–47

Accession number

P02679; P04469; P04470

Primary structure: γ chain

Sequence conflict: 114 I to K

γA variant:

Ala	A	28	Cys	C	11	Asp	D	32	Glu	E	22
Phe	F	20	Gly	G	35	His	H	11	Ile	I	27
Lys	K	33	Leu	L	32	Met	M	9	Asn	N	24
Pro	P	12	Gln	Q	24	Arg	R	11	Ser	S	29
Thr	T	29	Val	V	15	Trp	W	11	Tyr	Y	22

Mol. wt (calc.) = 49 426   Residues = 437

γB variant:

Ala	A	28	Cys	C	11	Asp	D	34	Glu	E	26
Phe	F	20	Gly	G	34	His	H	12	Ile	I	27
Lys	K	33	Leu	L	34	Met	M	9	Asn	N	24

Pro	P	15		Gln	Q	24		Arg	R	12		Ser	S	30
Thr	T	30		Val	V	15		Trp	W	11		Tyr	Y	24

Mol. wt (calc.) = 51 439   Residues = 453

```
1    MSWSLHPRNL  ILYFYALLFL  SSTCVAYVAT  RDNCCILDER  FGSYCPTTCG
51   IADFLSTYQT  KVDKDLQSLE  DILHQVENKT  SEVKQLIKAI  QLTYNPDESS
101  KPNMIDAATL  KSRIMLEEIM  KYEASILTHD  SSIRYLQEIY  NSNNQKIVNL
151  KEKVAQLEAQ  CQEPCKDTVQ  IHDITGKDCQ  DIANKGAKQS  GLYFIKPLKA
201  NQQFLVYCEI  DGSGNGWTVF  QKRLDGSVDF  KKNWIQYKEG  FGHLSPTGTT
251  EFWLGNEKIH  LISTQSAIPY  ALRVELEDWN  GRTSTADYAM  FKVGPEADKY
301  RLTYAYFAGG  DAGDAFDGFD  FGDDPSDKFF  TSHNGMQFST  WDNDNDKFEG
351  NCAEQDGSGW  WMNKCHAGHL  NGVYYQGGTY  SKASTPNGYD  NGIIWATWKT
401  RWYSMKKTTM  KIIPFNRLTI  GEGQQHHLGG  AKQ
```

γA variant continues:
```
434                                     AGDV
```

γB variant continues:
```
434                                     VRPEHPA     ETEYDSLYPE
451  DDL
```

Structural and functional sites
Signal peptide: 1–26
Potential N-linked glycosylation site: 78
Calcium-binding sites: 341, 355
Possible polymerization sites: 400, 422
QAGDV cell adhesion site: 433–437 (in γA variant only)
Cross-linking sites: 424, 432

Gene structure

Single copies of the genes for all three chains are closely linked on human chromosome 4. Two types of γ chain mRNAs can be produced in the human by alternative splicing [9].

References

[1] Crabtree, G.R. and Kant, J.A. (1981) Molecular cloning of cDNA for the α, β and γ chains of rat fibrinogen. J. Biol. Chem. 257: 7277–7279.

[2] Kant, J.A. et al (1983) Partial mRNA sequences for human Aα, Bβ and γ-fibrinogen chains: Evolutionary and functional implications. Proc. Natl Acad. Sci. USA 80: 3953–3957.

[3] Doolittle, R.F. (1984) Fibrinogen and fibrin. Annu. Rev. Biochem. 53: 195–229.

[4] Kloczewiak, M. et al (1984) Platelet receptor recognition site on human fibrinogen. Synthesis and structure-function relationship of peptides corresponding to the carboxy-terminal segment of the gamma chain. Biochemistry 23: 1767–1774.

[5] Kimura, S. and Aoki, N. (1986) Cross-linking site in fibrinogen for α2-plasmin inhibitor. J. Biol. Chem. 261: 15591–15595.

[6] Doolittle, R.F. (1987) Fibrinogen and fibrin. In: Haemostasis and Thrombosis, 2nd edition, Bloom, A.L. and Thomas, D.P., eds, Churchill Livingstone, Edinburgh, pp. 192–215.

[7] Doolittle, R.F. et al (1967) Amino acid sequence studies on artiodactyl fibrinopeptides. Arch. Biochem. Biophys. 118: 456–467.

[8] Doolittle, R.F. et al (1977) Amino acid sequence studies on the α chain of human fibrinogen. Characterization of 11 cyanogen bromide fragments. Biochemistry 16: 1703–1709.

[9] Oliasen, B. et al (1982) Fibrinogen gamma chain locus is on chromosome 4 in man. Hum. Genet. 61: 24–26.

Fibromodulin

Fibromodulin is a member of the small CS/DS proteoglycans group and has the same domain structure as decorin and biglycan. It was originally described as a 59 kDa connective tissue protein in cartilage and subsequently renamed to describe its ability to modulate collagen fibre formation. It is present in most tissues, including tendon, skin, sclera and cornea, but is somewhat more abundant in cartilage where it represents 0.1–0.3% of tissue wet weight. Like decorin, fibromodulin binds types I and II collagen fibrils *in vitro* and inhibits collagen fibril assembly. Fibromodulin is substituted with keratan sulphate GAG chains and has an NH_2-terminal tail of sulphated tyrosyl residues.

Molecular structure

Fibromodulin has a characteristic amino acid composition, with 14% of its residues being made up of leucine. Most of the protein consists of 10 repeats of 23 residues. The leucine-rich sequence shares homology with sequences in other proteins including decorin, biglycan, the serum protein LRG, platelet surface protein GPIb, ribonuclease/angiotensin inhibitor, chaoptin, toll protein and adenylate cyclase. Decorin, biglycan and fibromodulin are similar in size and the cysteine residues are located in conserved positions. The core protein is substituted with keratan sulphate side-chains and contains sulphated tyrosine [1–3].

Isolation

Fibromodulin can be isolated from cartilage by 4 M guanidine–HCl extraction and purified by ion-exchange chromatography and gel filtration [1].

Accession number

P13605

Primary structure

Ala A 16	Cys C 7	Asp D 18	Glu E 22
Phe F 13	Gly G 16	His H 11	Ile I 15
Lys K 12	Leu L 54	Met M 6	Asn N 30
Pro P 29	Gln Q 19	Arg R 18	Ser S 36
Thr T 11	Val V 16	Trp W 5	Tyr Y 21

Mol. wt (calc.) = 42 885   Residues = 375

```
  1    MQWASILLLA   GLCSLSWAQY   EEDSHWWFQF   LRNQQSTYDD   PYDPYPYEPY
 51    EPYPTGEEGP   AYAYGSPPQP   EPRDCPQECD   CPPNFPTAMY   CDNRNLKYLP
101    FVPSRMKYVY   FQNNQISSIQ   EGVFDNATGL   LWIALHGNQI   TSDKVGKKVF
151    SKLRHLERLY   LDHNNLTRIP   SPLPRSLREL   HLDHNQISRV   PNNALEGLEN
201    LTALYLHHNE   IQEVGSSMKG   LRSLILLDLS   YNHLRKVPDG   LPSALEQLYL
251    EHNNVFSVPD   SYFRGSPKLL   YVRLSHNSLT   NNGLASNTFN   SSSLLELDLS
```

```
301   YNQLQKIPPV   STNLENLYLQ   GNRINEFSIS   SFCTVVDVMN   FSKLQVQRLD
351   GNEIKRSAMP   ADAPLCLRLA   SLIEI
```

Structural and functional sites

Signal peptide: 1–18

Leucine-rich repeats: 114–137, 138–163, 164–184, 185–208, 209–231, 232–252, 253–276, 277–301, 302–321, 322–344

Potential N-linked glycosylation sites: 126, 165, 200, 290, 340

Potential sites for tyrosine sulphation: 20, 28, 32, 35, 37, 40, 43

Gene structure

The human fibromodulin gene spans approximately 8.5 kb.

References

[1] Heinegård, D. et al (1986) Two novel matrix proteins isolated from articular cartilage show wide distributions among connective tissues. J. Biol. Chem. 261: 3866–3872.

[2] Hedbom, E. and Heinegård, D. (1989) Interaction of a 59-kDa connective tissue matrix protein with collagen I and collagen II. J. Biol. Chem. 264: 6898–6905.

[3] Oldberg, Å. et al (1989) A collagen-binding 59-kd protein (fibromodulin) is structurally related to the small interstitial proteoglycans PG-S1 and PG-S2 (decorin). EMBO J. 8: 2601–2604.

Fibronectin

Fibronectin is a widely distributed glycoprotein present at high concentrations in most extracellular matrices, in plasma (300 μg/ml), and in other body fluids. Fibronectin is a prominent adhesive protein and mediates various aspects of cellular interactions with extracellular matrices including migration. Its principal functions appear to be in cellular migration during development and wound healing, regulation of cell growth and differentiation, and haemostasis/thrombosis.

Molecular structure

Fibronectin is a dimer of two non-identical subunits covalently linked near their COOH-termini by a pair of disulphide bonds. The difference between the subunits is determined by alternative splicing of the IIICS (or V) region. In the insoluble, matrix form of fibronectin, the dimer associates into disulphide-bonded oligomers and fibrils, while soluble, body fluid fibronectin is predominantly dimeric. Three regions of fibronectin are subject to alternative splicing and in general the matrix form of the molecule has a higher content of these segments than the soluble form. The human IIICS region has five potential variations, while the rat, bovine and chicken sequences have three, three and two, respectively. Each subunit is composed of a series of structurally independent domains linked by flexible polypeptide segments. At the primary sequence level, the origin of the majority of the fibronectin molecule can be accounted for by endoduplication of three types of polypeptide repeat. Different fibronectin domains are specialized for binding extracellular matrix macromolecules or bacterial or eukaryotic membrane receptors. The central cell-binding domain is recognized by most adherent cells via the integrin receptors α3β1, α5β1, αVβ1, αIIbβ3, αVβ3, αVβ5 and αVβ6. The IIICS/HepII cell-binding domain is recognized by lymphoid cells, neural crest derivatives and myoblasts via the integrins α4β1 and α4β7. Several peptide active sites have been identified in these domains [1–9].

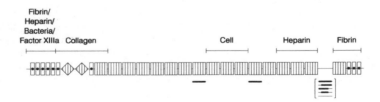

Isolation

Plasma fibronectin can be purified by a combination of gelatin and heparin affinity chromatography [10,11]. Cell-associated fibronectin can be extracted from culture monolayers with 1 M urea [11].

Accession number

P02751

Primary structure

Ala	A	100	Cys	C	63	Asp	D	126	Glu	E	145
Phe	F	54	Gly	G	208	His	H	51	Ile	I	121
Lys	K	78	Leu	L	136	Met	M	27	Asn	N	101
Pro	P	195	Gln	Q	133	Arg	R	126	Ser	S	200
Thr	T	268	Val	V	200	Trp	W	40	Tyr	Y	105

Mol. wt (calc.) = 273 715   Residues = 2476

```
1      MLRGPGPGLL  LLAVQCLGTA  VPSTGASKSK  RQAQQMVQPQ  SPVAVSQSKP
51     GCYDNGKHYQ  INQQWERTYL  GNVLVCTCYG  GSRGFNCESK  PEAEETCFDK
101    YTGNTYRVGD  TYERPKDSMI  WDCTCIGAGR  GRISCTIANR  CHEGGQSYKI
151    GDTWRRPHET  GGYMLECVCL  GNGKGEWTCK  PIAEKCFDHA  AGTSYVVGET
201    WEKPYQGWMM  VDCTCLGEGS  GRITCTSRNR  CNDQDTRTSY  RIGDTWSKKD
251    NRGNLLQCIC  TGNGRGEWKC  ERHTSVQTTS  SGSGPFTDVR  AAVYQPQPHP
301    QPPPYGHCVT  DSGVVYSVGM  QWLKTQGNKQ  MLCTCLGNGV  SCQETAVTQT
351    YGGNSNGEPC  VLPFTYNGRT  FYSCTTEGRQ  DGHLWCSTTS  NYEQDQKYSF
401    CTDHTVLVQT  QGGNSNGALC  HFPFLYNNHN  YTDCTSEGRR  DNMKWCGTTQ
451    NYDADQKFGF  CPMAAHEEIC  TTNEGVMYRI  GDQWDKQHDM  GHMMRCTCVG
501    NGRGEWTCYA  YSQLRDQCIV  DDITYNVNDT  FHKRHEEGHM  LNCTCFGQGR
551    GRWKCDPVDQ  CQDSETGTFY  QIGDSWEKYV  HGVRYQCYCY  GRGIGEWHCQ
601    PLQTYPSSSG  PVEVFITETP  SQPNSHPIQW  NAPQPSHISK  YILRWRPKNS
651    VGRWKEATIP  GHLNSYTIKG  LKPGVVYEGQ  LISIQQYGHQ  EVTRFDFTTT
701    STSTPVTSNT  VTGETTPFSP  LVATSESVTE  ITASSFVVSW  VSASDTVSGF
751    RVEYELSEEG  DEPQYLDLPS  TATSVNIPDL  LPGRKYIVNV  YQISEDGEQS
801    LILSTSQTTA  PDAPPDPTVD  QVDDTSIVVR  WSRPQAPITG  YRIVYSPSVE
851    GSSTELNLPE  TANSVTLSDL  QPGVQYNITI  YAVEENQEST  PVVIQQETTG
901    TPRSDTVPSP  RDLQFVEVTD  VKVTIMWTPP  ESAVTGYRVD  VIPVNLPGEH
951    GQRLPISRNT  FAEVTGLSPG  VTYYFKVFAV  SHGRESKPLT  AQQTTKLDAP
1001   TNLQFVNETD  STVLVRWTPP  RAQITGYRLT  VGLTRRGQPR  QYNVGPSVSK
1051   YPLRNLQPAS  EYTVSLVAIK  GNQESPKATG  VFTTLQPGSS  IPPYNTEVTE
1101   TTIVITWTPA  PRIGFKLGVR  PSQGGEAPRE  VTSDSGSIVV  SGLTPGVEYV
1151   YTIQVLRDGQ  ERDAPIVNKV  VTPLSPPTNL  HLEANPDTGV  LTVSWERSTT
1201   PDITGYRITT  TPTNGQQGNS  LEEVVHADQS  SCTFDNLSPG  LEYNVSVYTV
1251   KDDKESVPIS  DTIIPEVPQL  TDLSFVDITD  SSIGLRWTPL  NSSTIIGYRI
1301   TVVAAGEGIP  IFEDFVDSSV  GYYTVTGLEP  GIDYDISVIT  LINGGESAPT
1351   TLTQQTAVPP  PTDLRFTNIG  PDTMRVTWAP  PPSIDLTNFL  VRYSPVKNEE
1401   DVAELSISPS  DNAVVLTNLL  PGTEYVVSVS  SVYEQHESTP  LRGRQKTGLD
1451   SPTGIDFSDI  TANSFTVHWI  APRATITGYR  IRHHPEHFSG  RPREDRVPHS
1501   RNSITLTNLT  PGTEYVVSIV  ALNGREESPL  LIGQQSTVSD  VPRDLEVVAA
1551   TPTSLLISWD  APAVTVRYYR  ITYGETGGNS  PVQEFTVPGS  KSTATISGLK
1601   PGVDYTITVY  AVTGRGDSPA  SSKPISINYR  TEIDKPSQMQ  VTDVQDNSIS
1651   VKWLPSSSPV  TGYRVTTTPK  NGPGPTKTKT  AGPDQTEMTI  EGLQPTVEYV
1701   VSVYAQNPSG  ESQPLVQTAV  TNIDRPKGLA  FTDVDVDSIK  IAWESPQGQV
1751   SRYRVTYSSP  EDGIHELFPA  PDGEEDTAEL  QGLRPGSEYT  VSVVALHDDM
1801   ESQPLIGTQS  TAIPAPTDLK  FTQVTPTSLS  AQWTPPNVQL  TGYRVRVTPK
1851   EKTGPMKEIN  LAPDSSSVVV  SGLMVATKYE  VSVYALKDTL  TSRPAQGVVT
1901   TLENVSPPRR  ARVTDATETT  ITISWRTKTE  TITGFQVDAV  PANGQTPIQR
1951   TIKPDVRSYT  ITGLQPGTDY  KIYLYTLNDN  ARSSPVVIDA  STAIDAPSNL
2001   RFLATTPNSL  LVSWQPPRAR  ITGYIIKYEK  PGSPPREVVP  RPRPGVTEAT
2051   ITGLEPGTEY  TIYVIALKNN  QKSEPLIGRK  KTDELPQLVT  LPHPNLHGPE
```

```
2101   ILDVPSTVQK   TPFVTHPGYD   TGNGIQLPGT   SGQQPSVGQQ   MIFEEHGFRR
2151   TTPPTTATPI   RHRPRPYPPN   VGEEIQIGHI   PREDVDYHLY   PHGPGLNPNA
2201   STGQEALSQT   TISWAPFQDT   SEYIISCHPV   GTDEEPLQFR   VPGTSTSATL
2251   TGLTRGATYN   IIVEALKDQQ   RHKVREEVVT   VGNSVNEGLN   QPTDDSCFDP
2301   YTVSHYAVGD   EWERMSESGF   KLLCQCLGFG   SGHFRCDSSR   WCHDNGVNYK
2351   IGEKWDRQGE   NGQMMSCTCL   GNGKGEFKCD   PHEATCYDDG   KTYHVGEQWQ
2401   KEYLGAICSC   TCFGGQRGWR   CDNCRRPGGE   PSPEGTTGQS   YNQYSQRYHQ
2451   RTNTNVNCPI   ECFMPLDVQA   DREDSRE
```

Structural sites

Signal peptide: 1–20

Propeptide: 21–31

Type I repeats: 52–96, 97–140, 141–185, 186–230, 231–272, 308–344, 470–517, 518–560, 561–608, 2297–2341, 2342–2385, 2386–2428

Type II repeats: 345–404, 405–469

Type III repeats: 609–700, 719–809, 810–905, 906–995, 996–1085, 1086–1172, 1173–1265, 1357–1447, 1448–1537, 1538–1631, 1632–1721, 1812–1903, 1904–1992, 1993–2082, 2203–2273

Alternatively spliced domains: 1722–1811 (ED-A), 1266–1356 (ED-B), 2083–2202 (IIICS)

Potential N-linked glycosylation sites: 430, 528, 542, 877, 1007, 1244, 1291, 1904, 2199

O-Linked glycosylation site: 2155

Interchain disulphide bond residues: 2458, 2462

RGD cell adhesion site: 1615–1618

IDAPS cell adhesion site: 1994–1998

LDV cell adhesion site: 2102–2104

REDV cell adhesion site: 2182–2185

Heparin-binding sites: 2028–2046 (FN-C/H I), 2068–2082 (FN-C/H II)

Factor XIIIa transglutaminase cross-linking site: 34

Gene structure

Fibronectin is encoded by a single gene found on human chromosome 2 (at locus p14–16 or q34–36). The human gene is not fully characterized but contains approximately 50 exons. The chicken and rat genes span approximately 50 kb and 70 kb, respectively. Type I and II repeats correspond to single exons, type III repeats to two exons (except for III9). ED-A and ED-B are single exons. The IIICS is combined with the first half of III15 in one exon.

References

1 Pierschbacher, M.D. and Ruoslahti, E. (1984) The cell attachment activity of fibronectin can be duplicated by small synthetic fragments of the molecule. Nature 309: 30–33.

2 Yamada, K.M. and Kennedy, D.W. (1984) Dualistic nature of adhesive protein function: Fibronectin and its biologically active peptide fragments can autoinhibit fibronectin function. J. Cell Biol. 99: 29–36.

3 Kornblihtt, A.R. et al (1985) Primary structure of human fibronectin: Differential

splicing may generate at least 10 polypeptides from a single gene. EMBO J. 4: 1755–1759 .

4 Humphries, M.J. et al (1986) Identification of an alternatively spliced site in human plasma fibronectin that mediates cell type-specific adhesion. J. Cell Biol. 103: 2637–2647.

5 Humphries, M.J. et al (1987) Identification of two distinct regions of the type III connecting segment of human plasma fibronectin that promote cell type-specific adhesion. J. Biol. Chem. 262: 6886–6892.

6 McCarthy, J.B. et al (1988) Localization and chemical synthesis of fibronectin peptides with melanoma adhesion and heparin binding activities. Biochemistry 27: 1380–1388.

7 Mosher, D.F. (1989) Fibronectin, Academic Press, New York.

8 Hynes, R.O. (1990) Fibronectins, Springer-Verlag, New York.

9 Mould, A.P. and Humphries, M.J. (1991) Identification of a novel recognition sequence for the integrin α4β1 in the COOH-terminal heparin-binding domain of fibronectin. EMBO J. 10: 4089–4095.

10 Engvall, E. and Ruoslahti, E. (1977) Binding of soluble form of fibroblast surface protein, fibronectin, to collagen. Int. J. Cancer 20: 1–5.

11 Yamada, K.M. (1983) Isolation of fibronectin from plasma and cells. In: Immunochemistry of the Extracellular Matrix, Furthmayr, H., ed., CRC Press, Boca Raton, FL, pp. 111–123.

Fibulin

Fibulin is a glycoprotein found in the extracellular matrix and at moderate concentration in plasma (33 µg/ml). Initially, fibulin was suspected to be an intracellular protein interacting with the cytoplasmic domain of β1 integrins. Subsequent work showed it to be secreted by fibroblasts and incorporated into the extracellular matrix in a similar fashion to fibronectin. Apart from its calcium- and fibronectin-binding capacity, little is known about its role in the extracellular matrix or plasma.

Molecular structure

Fibulin is a single-chain polypeptide (calculated molecular weight of approximately 71 500), rich in cysteine (11%), which probably contains both N- and O-linked oligosaccharides. cDNA cloning indicates that three forms of fibulin exist, encoded by three transcripts that are likely to be derived from a common precursor mRNA. Fibulin can self-associate and bind to fibronectin. The NH_2-terminal portion of the molecule contains three repeated motifs that have potential disulphide loop structure and resemble the complement component anaphylatoxins C3a, C4a and C5a, as well as members of the albumin gene family. The bulk of the remainder of the molecule consists of nine cysteine-containing EGF repeats. Separating these two repeat domains is a 33 residue segment containing 12 acidic amino acids [1-5].

Isolation

Fibulin has been purified from placental extracts by affinity chromatography on a synthetic integrin β1 subunit cytoplasmic domain peptide–Sepharose column [1] or by monoclonal antibody affinity chromatography [4].

Accession number

P23142; P23143; P23144

Primary structure

Sequence conflicts: 36 C to S
41–42 HR to SH

A variant:

Ala	A	30	Cys	C	69	Asp	D	34	Glu	E	43
Phe	F	14	Gly	G	53	His	H	14	Ile	I	24
Lys	K	13	Leu	L	39	Met	M	5	Asn	N	26
Pro	P	23	Gln	Q	28	Arg	R	34	Ser	S	46
Thr	T	24	Val	V	28	Trp	W	0	Tyr	Y	19

Mol. wt (calc.) = 61 525   Residues = 566

B variant:

Ala	A	32	Cys	C	70	Asp	D	37	Glu	E	45
Phe	F	14	Gly	G	56	His	H	15	Ile	I	24
Lys	K	18	Leu	L	42	Met	M	5	Asn	N	27
Pro	P	25	Gln	Q	31	Arg	R	36	Ser	S	49
Thr	T	26	Val	V	29	Trp	W	1	Tyr	Y	19

Mol. wt (calc.) = 65 412   Residues = 601

C variant:

Ala	A	37	Cys	C	72	Asp	D	37	Glu	E	50
Phe	F	21	Gly	G	60	His	H	18	Ile	I	28
Lys	K	18	Leu	L	51	Met	M	8	Asn	N	29
Pro	P	33	Gln	Q	30	Arg	R	42	Ser	S	57
Thr	T	32	Val	V	39	Trp	W	0	Tyr	Y	21

Mol. wt (calc.) = 74 392   Residues = 683

A variant:

```
1     MERAAPSRRV   PLPLLLLGGL   ALLAAGVDAD   VLLEACCADG   HRMATHQKDC
51    SLPYATESKE   CRMVQEQCCH   SQLEELHCAT   GISLANEQDR   CATPHGDNAS
101   LEATFVKRCC   HCCLLGRAAQ   AQGQSCEYSL   MVGYQCGQVF   RACCVKSQET
151   GDLDVGGLQE   TDKIIEVEEE   QEDPYLNDRC   RGGGPCKQQC   RDTGDEVVCS
201   CFVGYQLLSD   GVSCEDVNEC   ITGSHSCRLG   ESCINTVGSF   RCQRDSSCGT
251   GYELTEDNSC   KDIDECESGI   HNCLPDFICQ   NTLGSFRCRP   KLQCKSGFIQ
301   DALGNCIDIN   ECLSISAPCP   IGHTCINTEG   SYTCQKNVPN   CGRGYHLNEE
351   GTRCVDVDEC   APPAEPCGKG   HRCVNSPGSF   RCECKTGYYF   DGISRMCVDV
401   NECQRYPGRL   CGHKCENTLG   SYLCSCSVGF   RLSVDGRSCE   DINECSSSPC
451   SQECANVYGS   YQCYCRRGYQ   LSDVDGVTCE   DIDECALPTG   GHICSYRCIN
501   IPGSFQCSCP   SSGYRLAPNG   RNCQDIDECV   TGIHNCSINE   TCFNIQGAFR
551   CLAFECPENY   RRSAAT
```

B variant continues after A variant:
```
567                            QKSK  KGRQNTPAGS   SKEDCRVLPW   KQGLEDTHLD
601   A
```

C variant continues after A variant:
```
567                            RCER  LPCHENRECS   KLPLRITYYH   LSFPTNIQAP
601   AVVFRMGPSS   AVPGDSMQLA   ITGGNEEGFF   TTRKVSPHSG   VVALTKPVPE
651   PRDLLLTVKM   DLSRHGTVSS   FVAKLFIFVS   AEL
```

Structural and functional sites
Signal peptide: 1–29
Type I repeats: 36–76, 77–111, 112–144
EGF (6C) repeats: 179–219, 220–265, 266–311, 312–359, 360–402, 403–444, 445–485, 486–528, 529–566
Potential N-linked glycosylation sites: 98, 535, 539
Alternative splicing sites: 566

Gene structure

No information is available.

109

References

[1] Argraves, W.S. et al (1989) Fibulin, a novel protein that interacts with the fibronectin receptor β subunit cytoplasmic domain. Cell 58: 623–629.

[2] Argraves, W.S. et al (1990) Fibulin is an extracellular matrix and plasma glycoprotein with repeated domain structure. J. Cell Biol. 111: 3155–3164.

[3] Kluge, M. et al (1990) Characterization of a novel calcium-binding, 90-kDa glycoprotein (BM–90) shared by basement membranes and serum. Eur. J. Biochem. 193: 651–659.

[4] Balbona, K. et al (1992) Fibulin binds to itself and to the carboxyl-terminal heparin-binding region of fibronectin. J. Biol. Chem. 267: 20120–20125.

[5] Spence, S.G. et al (1992) Fibulin is localized at sites of epithelial-mesenchymal transition in the early avian embryo. Devel. Biol. 151: 473–484.

Laminins

Laminins are a family of large glycoproteins that are distributed ubiquitously in basement membranes. These molecules are multifunctional, performing key roles in development, differentiation and migration through their ability to interact with cells via cell-surface receptors, including the integrins $\alpha 6\beta 1$, $\alpha 1\beta 1$, $\alpha 2\beta 1$ and $\alpha 3\beta 1$, and with other basement membrane components such as type IV collagen, entactin/nidogen and heparan sulphate proteoglycan. Studies on proteolytically derived fragments of EHS laminin have localized binding sites for collagen, heparin and entactin/nidogen, and defined domains of the molecule responsible for distinct biological activities such as cell attachment and neurite outgrowth promotion.

Molecular structure

Laminins are composed of three genetically distinct chains that assemble into a cruciform molecule with one long arm and three short arms (molecular weight 600 000–800 000). The long arm is formed by the association of the three chains into a triple coiled-coil that is stabilized by interchain disulphide bridges. The three short arms represent the N-terminal regions of each chain. Laminin molecules contain one heavy chain of molecular weight 300 000–400 000 and two light chains of molecular weight 220 000. Two heavy chains (A and M [for merosin]) and three light chains (B1, B2 and S) have been cloned and sequenced. These assemble into four isomeric forms of laminin, namely A–B1–B2 (EHS laminin), A–S–B2, M–B1–B2 and M–S–B2. B2t, a truncated form of the B2 chain, has recently been described, but as yet the chain composition of the molecule into which it assembles has not been determined. Isomeric forms of laminin exhibit developmental stage- and tissue-specific patterns of expression based on mRNA and immunolocalization studies.

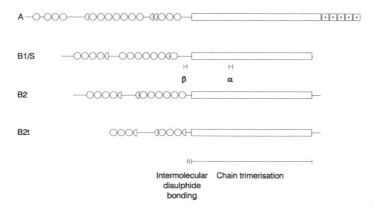

All laminin chains contain six domains. Domains I and II are located at the COOH-terminal ends of the light chains and in an equivalent position in the heavy chains. They consist of a series of heptad repeats with a predicted α-helical conformation which form the long arm of the assembled laminin molecule. Domains I and II are interrupted in the B1 and S chains by a short cysteine-rich domain (α). Domains III and V consist of homologous repeats of approximately 50 amino acid residues that are rich in glycine and contain eight cysteine residues arranged in a fashion that resembles the six-cysteine residue motif present in EGF and TGFα. A second domain III is present in the laminin A chain. Domains IV and VI are thought to form the globular domains in the short arms of the laminin molecule. The laminin A chain contains a second domain IV.

The laminin heavy chains (A and M) contain an extra domain at their COOH-terminus (domain G). The G domain consists of five homologous repeats (G1–G5) and is believed to form the large globular domain seen by electron microscopy at the end of the long arm. Laminins contain a variety of potential cell-binding sequences, although most of these are incompletely characterized. A non-conserved RGD site in the A chain interacts with the integrin αVβ3, the GD-6 peptide, again in the A chain, interacts with α3β1, and two B1 chain peptides, YIGSR and LGTIPG, probably interact with a 67 kDa binding protein [1-16].

Isolation

The most studied isoform of laminin (A–B1–B2) is obtained in large amounts by 0.5 M NaCl extraction of the mouse Engelbreth–Holm–Swarm (EHS) tumour [1] and is the only form of laminin available commercially. Alternatively, laminin associated with entactin/nidogen can be extracted from the EHS tumour and other tissues in a buffer containing 10 mM EDTA, 150 mM NaCl, 50 mM Tris–HCl, pH 7.4 in the presence of protease inhibitors [3].

Accession number

P25391

Primary structure: A chain

Sequence conflicts:
228–229	LQ	to	FE
252–254	IVT	to	MLP
419	H	to	E
519	V	to	L
1023	G	to	V
1075	D	to	V
1340	V	to	M
1513	R	to	P
2079–2080	NL	to	KV
2746	F	to	L
3054	F	to	L

Ala	A	193	Cys	C	166	Asp	D	171	Glu	E	187
Phe	F	86	Gly	G	256	His	H	91	Ile	I	122
Lys	K	141	Leu	L	289	Met	M	39	Asn	N	154
Pro	P	145	Gln	Q	132	Arg	R	161	Ser	S	244
Thr	T	173	Val	V	203	Trp	W	24	Tyr	Y	98

Mol. wt (calc.) = 336 777   Residues = 3075

```
1    MRGGVLLVLL   LCVAAQCRQR   GLFPAILNLA   SNAHISTNAT   CGEKGPEMFC
51   KLVEHVPGRP   VRNPQCRICD   GNSANPRERH   PISHAIDGTN   NWWQSPSIQN
101  GREYHWVTIT   LDLRQVFQVA   YVIIKAANAP   RPGNWILERS   LDGTTFSPWQ
151  YYAVSDSECL   SRYNITPRRG   PPTYRADDEV   ICTSYYSRLV   PLEHGEIHTS
201  LINGRPSADD   LSPKLLEFTS   ARYIRLRLQR   IRTLNADLMT   LSHREPKELD
```

251	PIVTRRYYYS	IKDISVGGMC	ICYGHASSCP	WDETTKKLQC	QCEHNTCGES
301	CNRCCPGYHQ	QPWRPGTVSS	GNTCEACNCH	NKAKDCYYDE	SVAKQKKSLN
351	TAGQFRGGGV	CINCLQNTMG	INCETCIDGY	YRPHKVSPYE	DEPCRPCNCD
401	PVGSLSSVCI	KDDLHSDLHN	GKQPGQCPCK	EGYTGEKCDR	CQLGYKDYPT
451	CVSCGCNPVG	SASDEPCTGP	CVCKENVEGK	ACDRCKPGFY	NLKEKNPRGC
501	SECFCFGVSD	VCSSLSWPVG	QVNSMSGWLV	TDLISPRKIP	SQQDALGGRH
551	QVSINNTAVM	QRLAPKYYWA	APEAYLGNKL	TAFGGFLKYT	VSYDIPVETV
601	DSNLMSHADV	IIKGNGLTLS	TQAEGLSLQP	YEEYLNVVRL	VPENFQDFHS
651	KRQIDRDQLM	TVLANVTHLL	IRATYNSAKM	ALYRLESVSL	DIASSNAIDL
701	VVAADVEHCE	CPQGYTGTSC	ESCLSGYYRV	DGILFGGICQ	PCECHGHAAE
751	CNVHGVCIAC	AHNTTGVHCE	QCLPGFYGEP	SRGTPGDCQP	CACPLTIASN
801	NFSPTCHLND	GDEVVCDWCA	PGYSGAWCER	CADGYYGNPT	VPGESCVPCD
851	CSGNVDPSEA	GHCDSVTGEC	LKCLGNTDGA	HCERCADGFY	GDAVTAKNCR
901	ACECHVKGSH	SAVCHLETGL	CDCKPNVTGQ	QCDQCLHGYY	GLDSGHGCRP
951	CNCSVAGSVS	DGCTDEGQCH	CVPGVAGKRC	DRCAHGFYAY	QDGSCTPCDC
1001	PHTQNTCDPE	TGECVCPPHT	QGGKCEECED	GHWGYDAEVG	CQACNCSLVG
1051	STHHRCDVVT	GHCQCKSKFG	GRACDQCSLG	YRDFPDCVPC	DCDLRGTSGD
1101	ACNLEQGLCG	CVEETGACPC	KENVFGPQCN	ECREGTFALR	ADNPLGCSPC
1151	FCSGLSHLCS	ELEDYVRTPV	TLGSDQPLLR	VVSQSNLRGT	TEGVYYQAPD
1201	FLLDAATVRQ	HIRAEPFYWR	LPQQFQGDQL	MAYGGKLKYS	VAFYSLDGVG
1251	TSNFEPQVLI	KGGRIRKQVI	YMDAPAPENG	VRQEQEVAMR	ENFWKYFNSV
1301	SEKPVTREDF	MSVLSDIEYI	LIKASYGQGL	QQSRISDISV	EVGRKAEKLH
1351	PEEEVASLLE	NCVCPPGTVG	FSCQDCAPGY	HRGKLPAGSD	RGPRPLVAPC
1401	VPCSCNNHSD	TCDPNTGKCL	NCGDNTAGDH	CDVCTSGYYG	KVTGSASDCA
1451	LCACPHSPPA	SFSPTCVLEG	DHDFRCDACL	LGYEGKHCER	CSSSYYGNPQ
1501	TPGGSCQKCD	CNRHGSVHGD	CDRTSGQCVC	RLGASGLRCD	ECEPRHILME
1551	TDCVSCDDEC	VGVLLNDLDE	IGDAVLSLNL	TGIIPVPYGI	LSNLENTTKY
1601	LQESLLKENM	QKDLGKIKLE	GVAEETDNLQ	KKLTRMLAST	QKVNRATERI
1651	FKESQDLAVA	IERLQMSITE	IMEKTTLNQT	LDEDFLLPNS	TLQNMQQNGT
1701	SLLEIMQIRD	FTQLHQNATL	ELKAAEDLLS	QIQENYQKPL	EELEVLKEAA
1751	SHVLSKHNNE	LKAAEALVRE	AEAKMQESNH	LLLMVNANLR	EFSDKKLHVQ
1801	EEQNLTSELI	VQGRGLIDAA	AAQTDAVQDA	LEHLEDHQDK	LLLWSAKIRH
1851	HIDDLVMHMS	QRNAVDLVYR	AEDHATEFQR	LADVLYSGLE	NIRNVSLNAT
1901	SAAYVHYNIQ	SLIEESEELA	RDAHRTVTET	SLLSESLVSN	GKAAVQRSSR
1951	FLKEGNNLSR	KLPGIALELS	ELRNKTNRFQ	ENAVEITRQT	NESLLILRAI
2001	PEGIRDKGAK	TKELATSASQ	SAVSTLRDVA	GLSQELLNTS	ASLSRVNTTL
2051	RETHQLLQDS	TMATLLAGRK	VKDVEIQANL	LFDRLKPLKM	LEENLSRNLS
2101	EIKLLISQAR	KQAASIKVAV	SADRDCIRAY	QPQISSTNYN	TLTLNVKTQE
2151	PDNLLFYLGS	STASDFLAVE	MRRGRVAFLW	DLGSGSTRLE	FPDFPIDDNR
2201	WHSIHVARFG	NIGSLSVKEM	SSNQKSPTKT	SKSPGTANVL	DVNNSTLMFV
2251	GGLGGQIKKS	PAVKVTHFKG	CLGEAFLNGK	SIGLWNYIER	EGKCRGCFGS
2301	SQNEDPSFHF	DGSGYSVVEK	SLPATVTQII	MLFNTFSPNG	LLLYLGSYGT
2351	KDFLSIELFR	GRVKVMTDLG	SGPITLLTDR	RYNNGTWYKI	AFQRNRKQGV
2401	LAVIDAYNTS	NKETKQGETP	GASSDLNRLD	KDPIYVGGLP	RSRVVRRGVT
2451	TKSFVGCIKN	LEISRSTFDL	LRNSYGVRKG	CLLEPIRSVS	FLKGGYIELP
2501	PKSLSPESEW	LVTFATTNSS	GIILAALGGD	VEKRGDREEA	HVPFFSVMLI
2551	GGNIEVHVNP	GDGTGLRKAL	LHAPTGTCSD	GQAHSISLVR	NRRIITVQLD
2601	ENNPVEMKLG	TLVESRTINV	SNLYVGGIPE	GEGTSLLTMR	RSFHGCIKNL
2651	IFNLELLDFN	SAVGHEQVDL	DTCWLSERPK	LAPDAEDSKL	LREPRAFPEQ
2701	CVVDAALEYV	PGAHQFGLTQ	NSHFILPFNQ	SAVRKKLSVE	LSIRTFASSG
2751	LIYYMAHQNQ	ADYAVLQLHG	GRLHFMFDLG	KGRTKVSHPA	LLSDGKWHTV
2801	KTDYVKRKGF	ITVDGRESPM	VTVVGDGTML	DVEGLFYLGG	LPSQYQARKI

113

```
2851  GNITHSIPAC  IGDVTVNSKQ  LDKDSPVSAF  TVNRCYAVAQ  EGTYFDGSGY
2901  AALVKEGYKV  QSDVNITLEF  RTSSQNGVLL  GISTAKVDAI  GLELVDGKVL
2951  FHVNNGAGRI  TPAYEPKTAT  VLCDGKWHTL  QANKSKHRIT  LIVDGNAVGA
3001  ESPHTQSTSV  DTNNPIYVGG  YPAGVKQKCL  RSQTSFRGCL  RKLALIKSPQ
3051  VQSFDFSRAF  ELHGVFLHSC  PGTES
```

Structural and functional sites
Signal peptide: 1–17

EGF (8C) repeats: 270–301, 327–394, 397–451, 454–500, 503–512 (partial), 709–739 (partial), 742–788, 791–846, 849–899, 902–948, 951–995, 998–1041, 1044–1087, 1090–1109 (partial), 1111–1147 (partial), 1150–1159 (partial), 1362–1400 (partial), 1403–1449, 1452–1506, 1509–1553

G repeats: 2140–2327, 2328–2509, 2510–2736, 2737–2913, 2914–3075

Potential N-linked glycosylation sites: 38, 164, 555, 665, 763, 801, 838, 926, 952, 1045, 1407, 1579, 1596, 1678, 1689, 1698, 1717, 1804, 1894, 1898, 1957, 1974, 1991, 2038, 2047, 2094, 2098, 2243, 2244, 2384, 2408, 2518, 2619, 2729, 2852, 2915, 2983

IKVAV cell adhesion site: 2116–2120

RGD cell adhesion site: 2534–2536 (not conserved in murine sequence)

GD–6 cell adhesion peptide: 3026–3047 (from analagous murine sequence)

Primary structure: M chain

Only limited sequence information is available from the COOH-terminal end of the human chain (Accession: P24043) [11].

Accession number
P07942

Primary structure: B1 chain

Sequence conflicts: 1470 L to V
                    1696 E to G

Ala	A	119	Cys	C	127	Asp	D	118	Glu	E	153
Phe	F	55	Gly	G	126	His	H	39	Ile	I	70
Lys	K	85	Leu	L	137	Met	M	32	Asn	N	80
Pro	P	87	Gln	Q	89	Arg	R	89	Ser	S	120
Thr	T	93	Val	V	98	Trp	W	13	Tyr	Y	56

Mol. wt (calc.) = 197 844   Residues = 1786

```
1    MGLLQLLAFS  FLALCRARVR  AQEPEFSYGC  AEGSCYPATG  DLLIGRAQKL
51   SVTSTCGLHK  PEPYCIVSHL  QEDKKCFICN  SQDPYHETLN  PDSHLIENVV
101  TTFAPNRLKI  WWQSENGVEN  VTIQLDLEAE  FHFTHLIMTF  KTFRPAAMLI
151  ERSSDFGKTW  GVYRYFAYDC  EASFPGISTG  PMKKVDDIIC  DSRYSDIEPS
201  TEGEVIFRAL  DPAFKIEDPY  SPRIQNLLKI  TNLRIKFVKL  HTLGDNLLDS
251  RMEIREKYYY  AVYDMVVRGN  CFCYGHASEC  APVDGFNEEV  EGMVHGHCMC
301  RHNTKGLNCE  LCMDFYHDLP  WRPAEGRNSN  ACKKCNCNEH  SISCHFDMAV
351  YLATGNVSGG  VCDDCQHNTM  GRNCEQCKPF  YYQHPERDIR  DPNFCERCTC
401  DPAGSQNEGI  CDSYTDFSTG  LIAGQCRCKL  NVEGEHCDVC  KEGFYDLSSE
451  DPFGCKSCAC  NPLGTIPGGN  PCDSETGHCY  CKRLVTGQHC  DQCLPEHWGL
```

```
 501   SNDLDGCRPC   DCDLGGALNN   SCFAESGQCS   CRPHMIGRQC   NEVEPGYYFA
 551   TLDHYLYEAE   EANLGPGVSI   VERQYIQDRI   PSWTGAGFVR   VPEGAYLEFF
 601   IDNIPYSMEY   DILIRYEPQL   PDHWEKAVIT   VQRPGRIPTS   SRCGNTIPDD
 651   DNQVVSLSPG   SRYVVLPRPV   CFEKGTNYTV   RLELPQYTSS   DSDVESPYTL
 701   IDSLVLMPYC   KSLDIFTVGG   SGDGVVTNSA   WETFQRYRCL   ENSRSVVKTP
 751   MTDVCRNIIF   SISALLHQTG   LACECDPQGS   LSSVCDPNGG   QCQCRPNVVG
 801   RTCNRCAPGT   FGFGPSGCKP   CECHLQGSVN   AFCNPVTGQC   HCFQGVYARQ
 851   CDRCLPGHWG   FPSCQPCQCN   GHADDCDPVT   GECLNCQDYT   MGHNCERCLA
 901   GYYGDPIIGS   GDHCRPCPCP   DGPDSGRQFA   RSCYQDPVTL   QLACVCDPGY
 951   IGSRCDDCAS   GYFGNPSEVG   GSCQPCQCHN   NIDTTDPEAC   DKETGRCLKC
1001   LYHTEGEHCQ   FCRFGYYGDA   LRQDCRKCVC   NYLGTVQEHC   NGSDCQCDKA
1051   TGQCLCLPNV   IGQNCDRCAP   NTWQLASGTG   CDPCNCNAAH   SFGPSCNEFT
1101   GQCQCMPGFG   GRTCSECQEL   FWGDPDVECR   ACDCDPRGIE   TPQCDQSTGQ
1151   CVCVEGVEGP   RCDKCTRGYS   GVFPDCTPCH   QCFALWDVII   AELTNRTHRF
1201   LEKAKALKIS   GVIGPYRETV   DSVERKVSEI   KDILAQSPAA   EPLKNIGNLF
1251   EEAEKLIKDV   TEMMAQVEVK   LSDTTSQSNS   TAKELDSLQT   EAESLDNTVK
1301   ELAEQLEFIK   NSDIRGALDS   ITKYFQMSLE   AEERVNASTT   EPNSTVEQSA
1351   LMRDRVEDVM   MERESQFKEK   QEEQARLLDE   LAGKLQSLDL   SAAAEMTCGT
1401   PPGASCSETE   CGGPNCRTDE   GERKCGGPGC   GGLVTVAHNA   WQKAMDLDQD
1451   VLSALAEVEQ   LSKMVSEAKL   RADEAKQSAE   DILLKTNATK   EKMDKSNEEL
1501   RNLIKQIRNF   LTQDSADLDS   IEAVANEVLK   MEMPSTPQQL   QNLTEDIRER
1551   VESLSQVEVI   LQHSAADIAR   AEMLLEEAKR   ASKSATDVKV   TADMVKEALE
1601   EAEKAQVAAE   KAIKQADEDI   QGTQNLLTSI   ESETAASEET   LFNASQRISE
1651   LERNVEELKR   KAAQNSGEAE   YIEKVVYTVK   QSAEDVKKTL   DGELDEKYKK
1701   VENLIAKKTE   ESADARRKAE   MLQNEAKTLL   AQANSKLQLL   KDLERKYEDN
1751   QRYLEDKAQE   LARLEGEVRS   LLKDISQKVA   VYSTCL
```

Structural and functional sites

Signal peptide: 1–21

EGF (8C) repeats: 271–334, 335–397, 398–457, 458–509, 510–540 (partial), 773–820, 821–866, 867–916, 917–975, 976–1027, 1028–1087, 1088–1121 (partial), 1122–1178

Potential N-linked glycosylation sites: 120, 356, 519, 677, 1041, 1195, 1279, 1336, 1343, 1487, 1542

LGTIPG cell adhesion site: 463–468

RYVVLPR (F9) cell adhesion site: 662–668

PDSGR cell adhesion site: 923–927

YIGSR cell adhesion site: 950–954

Accession number

P15800

Primary structure: S chain (rat)

Ala	A	166	Cys	C	125	Asp	D	96	Glu	E	113
Phe	F	46	Gly	G	171	His	H	69	Ile	I	39
Lys	K	33	Leu	L	161	Met	M	20	Asn	N	49
Pro	P	108	Gln	Q	117	Arg	R	154	Ser	S	107
Thr	T	90	Val	V	88	Trp	W	17	Tyr	Y	32

Mol. wt (calc.) = 196 253   Residues = 1801

```
   1   MEWASGKPGR  GRQGQPVPWE  LRLGLLLSVL  AATLAQVPSL  DVPGCSRGSC
  51   YPATGDLLVG  RADRLTASST  CGLHSPQPYC  IVSHLQDEKK  CFLCDSRRPF
 101   SARDNPNSHR  IQNVVTSFAP  QRRTAWWQSE  NGVPMVTIQL  DLEAEFHFTH
 151   LIMTFKTFRP  AAMLVERSAD  FGRTWRVYRY  FSYDCGADFP  GIPLAPPRRW
 201   DDVVCESRYS  EIEPSTEGEV  IYRVLDPAIP  IPDPYSSRIQ  NLLKITNLRV
 251   NLTRLHTLGD  NLLDPRREIR  EKYYYALYEL  VIRGNCFCYG  HASQCAPAPG
 301   APAHAEGMVH  GACICKHNTR  GLNCEQCQDF  YQDLPWHPAE  DGHTHACRKC
 351   ECNGHSHSCH  FDMAVYLASG  NVSGGVCDGC  QHNTAGRHCE  LCRPFFYRDP
 401   TKDMRDPAAC  RPCDCDPMGS  QDGGRCDSHD  DPVLGLVSGQ  CRCKEHVVGT
 451   RCQQCRDGFF  GLSASNPRGC  QRCQCNSRGT  VPGGTPCDSS  SGTCFCKRLV
 501   TGDGCDRCLP  GHWGLSHDLL  GCRPCDCDVG  GALDPQCDEA  TGQCPCRPHM
 551   IGRRCEQVQP  GYFRPFLDHL  TWEAEGAHGQ  VLEVVERLVT  NRETPSWTGV
 601   GFVRLREGQE  VEFLVTSLPR  AMDYDLLLRW  EPQVPEQWAE  LELVVQRPGP
 651   VSAHSPCGHV  LPRDDRIQGM  LHPNTRVLVF  PRPVCLEPGL  SYKLKLKLTG
 701   TGGRAHPETP  YSGSGILIDS  LVLQPHVLML  EMFSGGDAAA  LERRTTFERY
 751   RCHEEGLMPS  KTPLSEACVP  LLISASSLVY  NGALPCQCDP  QGSLSSECNP
 801   HGGQCRCKPG  VVGRRCDACA  TGYYGFGPAG  CQACQCSPDG  ALSALCEGTS
 851   GQCLCRTGAF  GLRCDHCQRG  QWGFPNCRPC  VCNGRADECD  AHTGACLGCR
 901   DYTGGEHCER  CIAGFHGDPR  LPYGGQCRPC  PCPEGPGSQR  HFATSCHRDG
 951   YSQQIVCHCR  AGYTGLRCEA  CAPGHFGDPS  KPGGRCQLCE  CSGNIDPTDP
1001   GACDPHTGQC  LRCLHHTEGP  HCGHCKPGFH  GQAARQSCHR  CTCNLLGTDP
1051   QRCPSTDLCH  CDPSTGQCPC  LPHVQGLSCD  RCAPNFWNFT  SGRGCQPCAC
1101   HPSRARGPTC  NEFTGQCHCH  AGFGGRTCSE  CQELHWGDPG  LQCRACDCDP
1151   RGIDKPQCHR  STGHCSCRPG  VSGVRCDQCA  RGFSGVFPAC  HPCHACFGDW
1201   DRVVQDLAAR  TRRLEQWAQE  LQQTGVLGAF  ESSFLNLQGK  LGMVQAIVAA
1251   RNTSAASTAK  LVEATEGLRH  EIGKTTERLT  QLEAELTDVQ  DENFNANHAL
1301   SGLERDGLAL  NLTLRQLDQH  LDILKHSNFL  GAYDSIRHAH  SQSTEAERRA
1351   NASTFAIPSP  VSNSADTRRR  AEVLMGAQRE  NFNRQHLANQ  QALGRLSTHT
1401   HTLSLTGVNE  LVCGAPGDAP  CATSPCGGAG  CRDEDGQPRC  GGLGCSGAAA
1451   TADLALGRAR  HTQAELQRAL  VEGGGILSRV  SETRRQAEEA  QQRAQAALDK
1501   ANASRGQVEQ  ANQELRELIQ  NVKDFLSQEG  ADPDSIEMVA  TRVLDISIPA
1551   SPEQIQRLAS  EIAERVRSLA  DVDTILAHTM  GDVRRAEQLL  QDAQRARSRA
1601   EGERQKAETV  QAALEEAQRA  QGAAQGAIRG  AVVDTKNTEQ  TLQQVQERMA
1651   GTEQSLNSAS  ERARQLHALL  EALKLKRAGN  SLAASTAEET  AGSAQSRARE
1701   AEKQLREQVG  DQYQTVRALA  ERKAEGVLAA  QARAEQLRDE  ARGLLQAAQD
1751   KLQRLQELEG  TYEENERELE  VKAAQLDGLE  ARMRSVLQAI  NLQVQIYNTC
1801   Q
```

Structural and functional sites
Signal peptide: 1–35
EGF (8C) repeats: 286–349, 350–412, 413–472, 473–524, 525–555 (partial), 786–833, 840–879, 880–929, 930–988, 989–1040, 1041–1081, 1098–1145, 1146–1192
Potential N-linked glycosylation sites: 251, 371, 1088, 1252, 1311, 1351, 1502
LRE cell adhesion site: 1705–1707

Accession number
P11047

Primary structure: B2 chain

Sequence conflict: 212 F to I

Ala	A	139	Cys	C	100	Asp	D	106	Glu	E	135
Phe	F	50	Gly	G	113	His	H	27	Ile	I	47
Lys	K	81	Leu	L	117	Met	M	22	Asn	N	100
Pro	P	72	Gln	Q	74	Arg	R	96	Ser	S	99
Thr	T	97	Val	V	78	Trp	W	11	Tyr	Y	45

Mol. wt (calc.) = 177 409   Residues = 1609

```
1      MRGSHRAAPA  LRPRGRLWPV  LAVLAAAAAA  GCAQAAMDEC  TDEGGRPQRC
51     MPEFVNAAFN  VTVVATNTCG  TPPEEYCVQT  GVTGVTKSCH  LCDAGQPHLQ
101    HGAAFLTDYN  NQADTTWWQS  QTMLAGVQYP  SSINLTLHLG  KAFDITYVRL
151    KFHTSRPESF  AIYKRTREDG  PWIPYQYYSG  SCENTYSKAN  RGFIRTGGDE
201    QQALCTDEFS  DFSPLTGGNV  AFSTLEGRPS  AYNFDNSPVL  QEWVTATDIR
251    VTLNRLNTFG  DEVFNDPKVL  KSYYYAISDF  AVGGRCKCNG  HASECMKNEF
301    DKLVCNCKHN  TYGVDCEKCL  PFFNDRPWRR  ATAESASECL  PCDCNGRSQE
351    CYFDPELYRS  TGHGGHCTNC  QDNTDGAHCE  RCRENFFRLG  NNEACSSCHC
401    SPVGSLSTQC  DSYGRCSCKP  GVMGDKCDRC  QPGFHSLTEA  GCRPCSCDPS
451    GSIDECNVET  GRCVCKDNVE  GFNCERCKPG  FFNLESSNPR  GCTPCFCFGH
501    SSVCTNAVGY  SVYSISSTFQ  IDEDGWRAEQ  RDGSEASLEW  SSERQDIAVI
551    SDSYFPRYFI  APAKFLGKQV  LSYGQNLSFS  FRVDRRDTRL  SAEDLVLEGA
601    GLRVSVPLIA  QGNSYPSETT  VKYVFRLHEA  TDYPWRPALT  PFEFQKLLNN
651    LTSIKIRGTY  SERSAGYLDD  VTLASARPGP  GVPATWVESC  TCPVGYGGQF
701    CEMCLSGYRR  ETPNLGPYSP  CVLCACNGHS  ETCDPETGVC  NCRDNTAGPH
751    CEKCSDGYYG  DSTAGTSSDC  QPCPCPGGSS  CAVVPKTKEV  VCTNCPTGTT
801    GKRCELCDDG  YFGDPLGRNG  PVRLCRLCQC  SDNIDPNAVG  NCNRLTGECL
851    KCIYNTAGFY  CDRCKDGFFG  NPLAPNPADK  CKACNCNPYG  TMKQQSSCNP
901    VTGQCECLPH  VTGQDCGACD  PGFYNLQSGQ  GCERCDCHAL  GSTNGQCDIR
951    TGQCECQPGI  TGQHCERCEV  NHFGFGPEGC  KPCDCHPEGS  LSLQCKDDGR
1001   CECREGFVGN  RCDQCEENYF  YNRSWPGCQE  CPACYRLVKD  KVADHRVKLQ
1051   ELESLIANLG  TGDEMVTDQA  FEDRLKEAER  EVMDLLREAQ  DVKDVDQNLM
1101   DRLQRVNNTL  SSQISRLQNI  RNTIEETGNL  AEQARAHVEN  TERLIEIASR
1151   ELEKAKVAAA  NVSVTQPEST  GDPNNMTLLA  EEARKLAERH  KQEADDIVRV
1201   AKTANDTSTE  AYNLLLRTLA  GENQTAFEIE  ELNRKYEQAK  NISQDLEKQA
1251   ARVHEEAKRA  GDKAVEIYAS  VAQLSPLDSE  TLENEANNIK  MEAENLEQLI
1301   DQKLKDYEDL  REDMRGKELE  VKNLLEKGKT  EQQTADQLLA  RADAAKALAE
1351   EAAKKGRDTL  QEANDILNNL  KDFDRRVNDN  KTAAEEALRK  IPAINQTITE
1401   ANEKTREAQQ  ALGSAAADAT  EAKNKAHEAE  RIASAVQKNA  TSTKAEAERT
1451   FAEVTDLDNE  VNNMLKQLQE  AEKELKRKQD  DADQDMMMAG  MASQAAQEAE
1501   INARKAKNSV  TSLLSIINDL  LEQLGQLDTV  DLNKLNEIEG  TLNKAKDEMK
1551   VSDLDRKVSD  LENEAKKQEA  AIMDYNRDIE  EIMKDIRNLE  DIRKTLPSGC
1601   FNTPSIEKP
```

Structural and functional sites

Signal peptide: 1–33

EGF (8C) repeats: 286–341, 342–397, 398–444, 445–494, 495–509 (partial), 690–723 (partial), 724–772, 773–827, 828–883, 884–934, 935–982, 983–1030

Potential N-linked glycosylation sites: 60, 134, 576, 650, 1022, 1107, 1161, 1175, 1205, 1223, 1241, 1380, 1395, 1439

RNIAEIIKDI (p20) cell adhesion site: 1577–1586 (from analagous murine sequence)

Accession number
Z15008

Primary structure: B2t chain

Ala	A	96	Cys	C	68	Asp	D	74	Glu	E	80
Phe	F	29	Gly	G	98	His	H	25	Ile	I	35
Lys	K	53	Leu	L	107	Met	M	21	Asn	N	57
Pro	P	56	Gln	Q	77	Arg	R	81	Ser	S	94
Thr	T	51	Val	V	55	Trp	W	6	Tyr	Y	30

Mol. wt (calc.) = 130 725   Residues = 1193

```
1      MPALWLGCCL  CFSLLLPAAR  ATSRREVCDC  NGKSRQCIFD  RELHRQTGNG
51     FRCLNCNDNT  DGIHCEKCKN  GFYRHRERDR  CLPCNCNSKG  SLSARCDNSG
101    RCSCKPGVTG  ARCDRCLPGF  HMLTDAGCTQ  DQRLLDSKCD  CDPAGIAGPC
151    DAGRCVCKPA  VTGERCDRCR  SGYYNLDGGN  PEGCTQCFCY  GHSASCRSSA
201    EYSVHKITST  FHQDVDGWKA  VQRNGSPAKL  QWSQRHQDVF  SSAQRLDPVY
251    FVAPAKFLGN  QQVSYGQSLS  FDYRVDRGGR  HPSAHDVILE  GAGLRITAPL
301    MPLGKTLPCG  LTKTYTFRLN  EHPSNNWSPQ  LSYFEYRRLL  RNLTALRIRA
351    TYGEYSTGYI  DNVTLISARP  VSGAPAPWVE  QCICPVGYKG  QFCQDCASGY
401    KRDSARLGPF  GTCIPCNCQG  GGACDPDTGD  CYSGDENPDI  ECADCPIGFY
451    NDPHDPRSCK  PCPCHNGFSC  SVIPETEEVV  CNNCPPGVTG  ARCELCADGY
501    FGDPFGEHGP  VRPCQPCQCN  SNVDPSASGN  CDRLTGRCLK  CIHNTAGIYC
551    DQCKAGYFGD  PLAPNPADKC  RACNCNPMGS  EPVGCRSDGT  CVCKPGFGGP
601    NCEHGAFSCP  ACYNQVKIQM  DQFMQQLQRM  EALISKAQGG  DGVVPDTELE
651    GRMQQAEQAL  QDILRDAQIS  EGASRSLGLQ  LAKVRSQENS  YQSRLDDLKM
701    TVERVRALGS  QYQNRVRDTH  RLITQMQLSL  AESEASLGNT  NIPASDHYVG
751    PNGFKSLAQE  ATRLAESHVE  SASNMEQLTR  ETEDYSKQAL  SLVRKALHEG
801    VGSGSGSPDG  AVVQGLVEKL  EKTKSLAQQL  TREATQAEIE  ADRSYQHSLR
851    LLDSVSPLQG  VSDQSFQVEE  AKRIKQKADS  LSSLVTRHMD  EFKRTQKNLG
901    NWKEEAQQLL  QNGKSGREKS  DQLLSRANLA  KSRAQEALSM  GNATFYEVES
951    ILKNLREFDL  QVDNRKAEAE  EAMKRLSYIS  QKVSDASDKT  QQAERALGSA
1001   AADAQRAKNG  AGEALEISSE  IEQEIGSLNL  EANVTADGAL  AMEKGLASLK
1051   SEMREVEGEL  ERKELEFDTN  MDAVQMVITE  AQKVDTRAKN  AGVTIQDTLN
1101   TLDGLLHLMD  QPLSVDEEGL  VLLEQKLSRA  KTQINSQLRP  MMSELEERAR
1151   QQRGHLHLLE  TSIDGILADV  KNLENIRDNL  PPGCYNTQAL  EQQ
```

Variant is identical up to 1109 then:
```
1110             G  M
```

Structural and functional sites
Signal peptide: 1–21
EGF (8C) repeats: 28–83, 84–130, 139–185, 186–196 (partial), 382–415 (partial), 416–461, 462–516, 517–572, 573–608
Potential N-linked glycosylation sites: 224, 326, 342, 362, 942, 1033

Gene structure

The laminin A chain gene is localized to human chromosome 18p11.3. The B1 chain contains 34 exons varying in size from 64 to 370 bp in a gene of >80 kb and is localized to chromosome 7q22. The B2 gene contains 28 exons varying in size from 80 to 3051 bp in a gene of >58 kb and is localized to chromosome 1q25–q31, an identical location to B2t. No information is available for either the S or M chain genes [5,13,16,17].

References

[1] Timpl, R. et al (1979) Laminin – a glycoprotein from basement membranes. J. Biol. Chem. 254: 9933–9937.

[2] Graf, J. et al (1987) Identification of an amino acid sequence in laminin mediating cell attachment, chemotaxis, and receptor binding. Cell 48: 989–996.

[3] Paulsson, M. et al (1987) Laminin-nidogen complex. Extraction with chelating agents and structural characterisation. Eur. J. Biochem. 166: 467–478.

[4] Pikkarainen, T. et al (1987) Human laminin B1 chain. J. Biol. Chem. 262: 10454–10462.

[5] Fukushima, Y. et al (1988) Isolation of a human laminin B2 (LAMB2) cDNA clone and assignment of the gene to chromosome region 1q25–q31. Cytogenet. Cell Genet. 48: 137–141.

[6] Pikkarainen, T. et al (1988) Human laminin B2 chain. J. Biol. Chem. 263: 6751–6758.

[7] Hunter, D.D. et al (1989) Primary sequence of a motor neuron-selective adhesive site in the synaptic basal lamina protein S-laminin. Cell 59: 905–913.

[8] Kleinman, H.K. et al (1989) Identification of a second active site in laminin for promotion of cell adhesion, migration, and inhibition of in vivo melanoma lung colonization. Arch. Biochem. Biophys. 272: 39–45.

[9] Liesi, P. et al (1989) Identification of a neurite outgrowth-promoting domain of laminin using synthetic peptides. FEBS Lett. 244: 141–148.

[10] Tashiro, K.-I. et al (1989) A synthetic peptide containing the IKVAV sequence from the A chain of laminin mediates cell attachment, migration, and neurite outgrowth. J. Biol. Chem. 264: 16174–16182.

[11] Ehrig, K. et al (1990) Merosin, a tissue-specific basement membrane protein, is a laminin-like protein. Proc. Natl Acad. Sci. USA 87: 3264–3268.

[12] Vuolteenaho, R. et al (1990) Structure of the human laminin B1 gene. J. Biol. Chem. 265: 15611–15616.

[13] Kallunki, T. et al (1991) Structure of the laminin B2 gene reveals extensive divergence from the laminin B1 chain gene. J. Biol. Chem. 266: 221–228.

[14] Nissinen, M. et al (1991) Primary structure of the human laminin A chain. Limited expression in human tissues. Biochem. J. 276: 369–379.

[15] Gehlsen, K.R. et al (1992) A synthetic peptide derived from the carboxy terminus of the laminin A chain represents a binding site for the α3β1 integrin. J. Cell Biol. 117: 449–459.

[16] Kallunki, P. et al (1992) A truncated laminin chain homologous to the B2 chain: Structure, spatial expression and chromosomal assignment. J. Cell Biol. 119: 679–693.

[17] Nagayoshi, T. et al (1989) Human laminin A chain (LAMA) gene: Chromosomal mapping to locus 18p11.3. Genomics 5: 932–935.

Link protein

Link protein binds to both cartilage proteoglycan (aggrecan) and hyaluronan in cartilage extracellular matrix, thereby stabilizing their aggregation and producing supramolecular assemblies. It is a glycoprotein of approximate molecular weight 45 000. Three forms have been identified: LP1 and LP2 differ in glycosylation at their NH2-termini, and LP3 is a biologically active proteolytic cleavage product that lacks 13–16 amino acids at its NH2-terminus. Link protein has been detected in other connective tissues and may also interact with proteoglycans (such as versican and neurocan) and hyaluronan in these locations.

Molecular structure

Link protein exhibits homology to the G1 domain of aggrecan and contains both an immunoglobulin repeat and a two-loop structure stabilized by disulphide bonds. The NH2-terminal portion of the molecule binds to the proteoglycan and has homology to an immunoglobulin variable repeat [1-6].

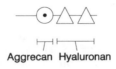

Aggrecan Hyaluronan

Isolation

A 4 M guanidine–HCl extract of cartilage is reassociated and separated by isopycnic caesium chloride density gradient centrifugation. The bottom fraction containing proteoglycan aggregates is taken, dissociated, recentrifuged and the top fraction containing link protein isolated. Further purification is achieved by gel filtration [3].

Accession number
P10915

Primary structure

Ala	A	25	Cys	C	11	Asp	D	29	Glu	E	12
Phe	F	18	Gly	G	31	His	H	10	Ile	I	16
Lys	K	20	Leu	L	31	Met	M	2	Asn	N	12
Pro	P	14	Gln	Q	13	Arg	R	21	Ser	S	18
Thr	T	15	Val	V	26	Trp	W	8	Tyr	Y	22

Mol. wt (calc.) = 40 120   Residues = 354

```
  1    MKSLLLLVLI   SICWADHLSD   NYTLDHDRAI   HIQAENGPHL   LVEAEQAKVF
 51    SHRGGNVTLP   CKFYRDPTAF   GSGIHKIRIK   WTKLTSDYLK   EVDVFVSMGY
101    HKKTYGGYQG   RVFLKGGSDS   DASLVITDLT   LEDYGRYKCE   VIEGLEDDTV
151    VVALDLQGVV   FPYFPRLGRY   NLNFHEAQQA   CLDQDAVIAS   FDQLYDAWRG
201    GLDWCNAGWL   SDGSVQYPIT   KPREPCGGQN   TVPGVRNYGF   WDKDKSRYDV
251    FCFTSNFNGR   FYYLIHPTKL   TYDEAVQACL   NDGAQIAKVG   QIFAAWKILG
```

```
301  YDRCDAGWLA  DGSVRYPISR  PRRRCSPTEA  AVRFVGFPDK  KHKLYGVYCF
351  RAYN
```

Structural and functional sites
Signal peptide: 1–15
Immunoglobulin repeat: 47–142
Link repeats: 176–253, 274–350
Potential N-linked glycosylation sites: 21, 56
Disulphide bonds: 61–139, 181–252, 205–226, 279–349, 304–325

Gene structure

A single proteoglycan link protein gene is found on chromosome 5 at locus q13–14 [5].

References
[1] Caterson, B. et al (1985) Monoclonal antibodies as probes for determining the microheterogeneity of the link proteins of cartilage proteoglycans. J. Biol. Chem. 260: 11348–11356.
[2] Doege, K. et al (1986) Link protein cDNA sequence reveals a tandemly repeated protein structure. Proc. Natl Acad. Sci. USA 83: 3761–3765 .
[3] Neame, P.J. et al (1986) The primary structure of link protein from rat chondrosarcoma proteoglycan aggregate. J. Biol. Chem. 261: 3519–3535.
[4] Goetinck, P.F. et al (1987) The tandemly repeated sequences of cartilage link protein contain the sites for interaction with hyaluronic acid. J. Cell Biol. 105: 2403–2408.
[5] Byers, M.G. et al (1990) Complete amino acid sequence of human cartilage link protein (CRTL1) deduced from cDNA clones and chromosomal assignment of the gene. Genomics 8: 562–567.
[6] Dudhia, J. and Hardingham, T.E. (1990) Primary structure of human cartilage-link protein. Nucleic Acids Res. 18: 1292.

Lumican   corneal keratan sulphate proteoglycan

Lumican is a small keratan sulphate proteoglycan whose core protein is homologous to decorin, biglycan and fibromodulin. It is expressed in a number of tissues including cornea (it was originally found in chicken cornea), muscle, aorta and intestine. Several lines of evidence suggest that it is important in corneal transparency: it is present early in the development of the chick cornea, it is absent from the opaque cornea of healing rabbit corneal wounds, and the return of transparency is accompanied by the presence of lumican. The abundance of low sulphate lumican in many tissues indicates that this protein occurs predominantly as a glycoprotein rather than the highly sulphated proteoglycan present in cornea.

Molecular structure

Lumican has a central leucine-rich domain that accounts for 62% of the protein. The central domain contains nine repeats of the sequence LXXLXLXXNXL/I where X is any amino acid. There are three variations of this sequence arranged in tandem to form three repeating units [1–4].

Isolation

Lumican can be purified from extracts of aorta using DEAE ion-exchange chromatography, gel filtration, hydrophobic interaction and reverse-phase chromatography [5].

Accession number

A41748

Primary structure

Ala	A	11	Cys	C	7	Asp	D	19	Glu	E	13
Phe	F	15	Gly	G	17	His	H	8	Ile	I	22
Lys	K	22	Leu	L	55	Met	M	5	Asn	N	34
Pro	P	20	Gln	Q	12	Arg	R	6	Ser	S	24
Thr	T	20	Val	V	16	Trp	W	1	Tyr	Y	16

Mol. wt (calc.) = 38 598   Residues = 343

```
1    MTLNSLPIFL   VLISGIFCQY   DYGPADDYGY   DPFGPSTAVC   APECNCPLSY
51   PTAMYCDNLK   LKTIPIVPSG   IKYLYLRNNM   IEAIEENTFD   NVTDLQWLIL
101  DHNHLENSKI   KGRVFSKLKN   LKKLHINYNN   LTEAVGPLPK   TLDDLQLSHN
151  KITKVNPGAL   EGLVNLTVIH   LQNNQLKTDS   ISGAFKGLNS   LLYLDLSFNQ
201  LTKLPTGLPH   SLLMLYFDNN   QISNIPDEYF   QGFKTLQYLR   LSHNKLTDSG
251  IPGNVFNITS   LVELDLSFNQ   LKSIPTVSEN   LENFYLQVNK   INKFPLSSFC
301  KVVGPLTYSK   ITHLRLDGNN   LTRADLPQEM   YNCLRVAADI   SLE
```

Structural and functional sites

Signal peptide: 1–18
Leucine-rich repeats: 67–137, 138–207, 208–277
Potential N-linked glycosylation sites: 91, 130, 165, 157, 120

Gene structure

Unknown.

References

[1] Cintron, C. et al (1978) Biochemical and ultrastructural changes in collagen during corneal wound healing. J. Ultrastruct. Res. 65: 13–22.

[2] Hassell, J.R. et al (1983) Proteoglycan changes during restoration of transparency in corneal scars. Arch. Biochem. Biophys. 222: 362–369.

[3] Funderburgh, J.L. et al (1986) Keratan sulfate proteoglycan during embryonic development of the chicken cornea. Devel. Biol. 116: 267–277.

[4] Blochberger, T.C. et al (1992) cDNA to chick lumican (corneal keratan sulfate proteoglycan) reveals homology to the small interstitial proteoglycan gene family and expression in muscle and intestine. J. Biol. Chem. 267: 347–352.

[5] Funderburgh, J.L et al (1991) Unique glycosylation of 3 keratan sulfate proteoglycan isoforms. J. Biol. Chem. 266: 14226–14231.

Matrix Gla protein

Matrix Gla protein is a vitamin K-dependent protein initially isolated from bovine bone and associated with the organic matrix. The protein was subsequently shown to be expressed in many tissues including cartilage and most visceral organs. In each tissue it is restricted to a discrete set of tissue-specific cells. In bone, but not in kidney, 1,25-dihydroxyvitamin D3 upregulates matrix Gla protein expression. Matrix Gla protein is the only known vitamin K-dependent protein that lacks a propeptide.

Molecular structure

Matrix Gla protein is a single-chain polypeptide which contains five Gla (γ-carboxyglutamic acid) residues and is stabilized by one intrachain disulphide bond. Despite its high content of hydrophilic amino acids, it is exceptionally water-insoluble and requires 4 M guanidine–HCl for extraction. The two forms of matrix Gla protein isolated so far contain 79 and 83 residues and lack 5 and 1 amino acids from the predicted COOH-terminus. Residues 2–12 encode an α-helical portion which shares no homology with other proteins. This is followed by a putative γ-carboxylation recognition site and a Gla-containing region. This region shows some identity to osteocalcin and other vitamin K-dependent proteins [1-6].

Isolation

Matrix Gla protein can be isolated by extracting demineralized bone matrix with denaturing agents and fractionating the extract by gel filtration and ion-exchange chromatography [1].

Accession number

P08493

Primary structure

Ala A 11	Cys C 3	Asp D 2	Glu E 9
Phe F 3	Gly G 2	His H 2	Ile I 5
Lys K 5	Leu L 9	Met M 3	Asn N 7
Pro P 3	Gln Q 3	Arg R 14	Ser S 6
Thr T 2	Val V 5	Trp W 1	Tyr Y 8

Mol. wt (calc.) = 12 309   Residues = 103

```
1     MKSLILLAIL   AALAVVTLCY   ESHESMESYE   LNPFINRRNA   NTFISPQQRW
51    RAKVQERIRE   RSKPVHELNR   EACDDYRLCE   RYAMVYGYNA   AYNRYFRKRR
101   GAK
```

Structural and functional sites

Signal peptide: 1–19
Intrachain disulphide bond: 73, 79
γ-Carboxyglutamic acid residues: 21, 56, 60, 67, 71

Gene structure

The human matrix Gla protein gene spans 3.9 kb of chromosomal DNA and consists of four exons separated by intervening sequences which account for more than 80% of the gene. The gene has been assigned to the short arm (p) of human chromosome 12. The four-exon organization is similar to that of osteocalcin, but different from the two-exon organization found in other vitamin K-dependent proteins [5].

References
[1] Price, P.A. and Williamson, M.K. (1985) Primary structure of bovine matrix Gla protein, a new vitamin K-dependent bone protein. J. Biol. Chem. 260: 14971–14975.
[2] Price, P.A. et al (1987) Molecular cloning of matrix Gla protein. Implications for substrate recognition by the vitamin K-dependent γ carboxylase. Proc. Natl Acad. Sci. USA 84: 8335–8339.
[3] Fraser, J.D. and Price, P.A. (1988) Lung, heart, and kidney express high levels of mRNA for the vitamin K-dependent matrix Gla protein. J. Biol. Chem. 263: 11033–11036.
[4] Kiefer, M.C. et al (1988) The cDNA and derived amino acid sequences of human and bovine matrix Gla protein. Nucleic Acids Res. 16: 5213.
[5] Cancela, L. et al (1990) Molecular structure, chromosome assignment, and promoter organisation of the human matrix Gla protein gene. J. Biol. Chem. 265: 15040–15048.
[6] Chen, L. et al (1990) Overexpression of matrix Gla protein mRNA in malignant human breast cells: Isolation by differential cDNA hybridization. Oncogene 5: 1391–1395.

Microfibril-associated glycoprotein

Microfibril-associated glycoprotein is a component of the 12 nm microfibrils that are found in a variety of elastic and non-elastic tissues. In elastic tissues, these "elastin-associated microfibrils" become incorporated into elastic fibres and act as elastic fibre determinants. Microfibril-associated glycoprotein has been shown to localize on the beaded microfibrils which contain fibrillin as a major component.

Molecular structure

Microfibril-associated glycoprotein is an acidic glycoprotein which is synthesized as a 21 kDa polypeptide. At least two structural regions have been described: an NH_2-terminal domain containing high levels of glutamine, proline and acidic amino acids, and a COOH-terminal domain which contains all 13 cysteine residues and most of the basic amino acids. The presence of high levels of proline and glutamic acid has been suggested to account for the anomalous electrophoretic behaviour of microfibril-associated glycoprotein. There appears to be no N-glycosylation, and O-linked sites to serine or threonine have not been delineated [1–5].

Isolation

Glycoproteins from elastin-associated microfibrils have been solubilized from foetal bovine nuchal ligament using treatment with reductive saline extracts. Microfibril-associated glycoprotein can be purified in low yield from this extract by DEAE chromatography followed by gel filtration on Sephacryl S–300 [4]. Microfibril-associated glycoprotein was originally prepared from reductive guanidinium–HCl extracts.

Accession number
M59851

Primary structure (bovine)

Ala A 12	Cys C 13	Asp D 8	Glu E 16
Phe F 6	Gly G 8	His H 3	Ile I 4
Lys K 5	Leu L 18	Met M 2	Asn N 3
Pro P 18	Gln Q 16	Arg R 10	Ser S 10
Thr T 8	Val V 12	Trp W 0	Tyr Y 11

Mol. wt (calc.) = 20 685   Residues = 183

```
1    MRAASLFLLF  LPAGLLAQGQ  YDLDPLPPYP  DHVQYTHYSE  QIENPDYYDY
51   PEMTPRPPEE  QFQFQSQQQV  QQEVIPAPTL  EPGTVETEPT  EPGPLDCREE
101  QYPCTRLYSI  HKPCKQCLNE  VCFYSLRRVY  VVNKEICVRT  VCAQEELLRA
151  DLCRDKFSKC  GVLASSGLCQ  SVAAACARSC  GGC
```

Structural and functional sites
Signal peptide: 1–17 (possibly 1–19)

Gene structure

Northern blot hybridization of poly(A) RNA has identified a single mRNA species of approximately 1.1 kb. Gene structure is unknown.

References

1 Gibson, M.A. et al (1986) The major antigen of elastin-associated microfibrils is a 31-kDa glycoprotein. J. Biol. Chem. 261: 11429–11436.

2 Cleary, E.G. (1987) The microfibrillar component of elastic fibres: Morphology and biochemistry. In: Connective Tissue Disease. Molecular Pathology of the Extracellular Matrix, Uitto, J. and Perjeda, A.J., eds, Marcel Dekker, New York, pp. 55–81.

3 Gibson, M.A. and Cleary, E.G. (1987) The immunohistochemical localisation of microfibril-associated glycoprotein (MAGP) in elastic and non-elastic tissues. Immunol. Cell Biol. 65: 345–356.

4 Gibson, M.A. et al (1989) The protein components of the 12 nm microfibrils of elastic and non-elastic tissues. J. Biol. Chem. 264: 4590–4598.

5 Gibson, M.A. et al (1991) Complementary DNA cloning establishes microfibril-associated glycoprotein (MAGP) to be a discrete component of the elastin-associated microfibrils. J. Biol. Chem. 266: 7596–7601.

Osteocalcin
gamma-carboxyglutamic acid-containing protein, bone Gla-protein

Osteocalcin is a low molecular weight protein found in bones, teeth and possibly other mineralized tissues. It is highly conserved among all vertebrate species studied and usually contains three γ-carboxyglutamic acid residues (the human protein may only contain two), which provide the molecule with its calcium-binding properties. Prior to mineralization, osteocalcin is found in only trace amounts, but its level increases markedly during periods of intense skeletal growth. Osteocalcin binds tightly to hydroxyapatite and may function in the assembly of mineralized bone by regulating apatite crystal growth. Its synthesis is vitamin K-dependent and is stimulated by 1,25-dihydroxyvitamin D3. Specific osteoblastic synthesis and presence in blood has led to its use as a diagnostic parameter of bone metabolism.

Molecular structure

Osteocalcin is secreted as a precursor (approximate molecular weight 10 000) and processed to the active molecule of 46–50 (approximate molecular weight 5800) residues stabilized by a single intrachain disulphide bond. The glutamic acid residues to be modified by vitamin K-dependent carboxylation are clustered in the centre of the molecule and are preceded by a putative γ-carboxylation recognition site. The amino acid sequence of this site shows identity with matrix Gla protein. A single hydroxyproline residue occurs in most species [1–6].

Propeptide

Isolation
Osteocalcin can be isolated from EDTA extracts of bone by gel filtration on Sephadex G-100 and DEAE–Sephadex ion-exchange chromatography [1]. Protein degradation can be avoided by demineralizing in 20% formic acid, fractionating on Sephacryl S-200 and immunoabsorbing [2].

Accession number
P02818

Primary structure
Sequence conflict: 33–34 missing

Ala	A	14	Cys	C	3	Asp	D	4	Glu	E	8
Phe	F	3	Gly	G	8	His	H	1	Ile	I	2
Lys	K	4	Leu	L	12	Met	M	1	Asn	N	1
Pro	P	9	Gln	Q	4	Arg	R	8	Ser	S	5
Thr	T	1	Val	V	6	Trp	W	1	Tyr	Y	5

Mol. wt (calc.) = 10 950   Residues = 100

```
 1    MRALTLLALL   ALAALCIAGQ   AGAKPSGAES   SKGAAFVSKQ   EGSEVVKRPR
51    RYLYQWLGAP   VPYPDPLEPR   REVCELNPDC   DELADHIGFQ   EAYRRFYGPV
```

Structural and functional sites
Signal peptide: 1–23
Propeptide: 24–51
γ-Carboxyglutamic acid residues: 68 (9% of cases), 72, 75
Intrachain disulphide bond: 74–80

Gene structure

The osteocalcin gene contains four exons. It has been mapped to the long arm of human chromosome 1 by Southern blot analysis [7].

References
[1] Price, P.A. et al (1976) Characterization of a γ-carboxyglutamic acid-containing protein from bone. Proc. Natl Acad. Sci. USA 73: 1447–1451.
[2] Poser, J.W. et al (1980) Isolation and sequence of the vitamin K-dependent protein from human bone. J. Biol. Chem. 255: 8685–8691.
[3] Pan, L.C. and Price, P.A. (1985) The propeptide of rat bone γ-carboxyglutamic acid protein shares homology with other vitamin K-dependent protein precursors. Proc. Natl Acad. Sci. USA 82: 6109–6113.
[4] Celeste, A.J. et al (1986) Isolation of the human gene for bone gla protein utilizing mouse and rat cDNA clones. EMBO J. 5: 1885–1890.
[5] Price, P.A. (1987) Vitamin K-dependent bone proteins. In: Calcium Regulation and Bone Metabolism. Basic and Clinical Aspects, Cohn D.V. et al, eds, Vol. 9, Elsevier, Amsterdam, pp. 419–425
[6] Kieffer, M.C. et al (1990) The cDNA and derived amino acid sequences of human and bovine Gla protein. Nucleic Acids Res. 18: 1909.
[7] Puchacz, E. et al (1989) Chromosomal localization of the human osteocalcin gene. Endocrinology 124: 2648–2650.

Osteonectin   BM40, SPARC

Osteonectin was initially isolated as one of the main non-collagenous components of bone. However, a much broader functional role is indicated by its expression in a variety of tissues including endodermal, epidermal and soft connective tissues. The protein contains a calcium-binding region, but its function is unknown. Expression of osteonectin has been linked to bone formvation and to tissue differentiation and remodelling due to an ability of some forms to bind collagens and hydroxyapatite.

Molecular structure

Osteonectin is a single-chain polypeptide composed of four distinct domains. The NH2-terminal domain (I) is encoded by exons 3 and 4, contains two glutamate-rich segments and can bind up to eight Ca^{2+} ions. Domain II is encoded by exons 5 and 6, shows some homology to ovomucoid and certain serine proteinase inhibitors, and is rich in intrachain disulphide bonds. Domain III, encoded by exons 7 and 8, is predicted to be α-helical. Exon 9 encodes domain IV and contains a single EF-hand-type, helix–loop–helix cation-binding structure. The protein is glycosylated and some forms may be phosphorylated [1–7].

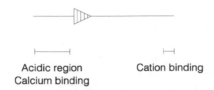

| Acidic region | Cation binding |
| Calcium binding | |

Isolation

Osteonectin has been isolated from both EDTA and guanidine–HCl/EDTA extracts of bone and can be purified by ion-exchange chromatography and gel permeation [1].

Accession number

P09486

Primary structure

Sequence conflict: 42–43 SC to PT

Ala	A	18	Cys	C	15	Asp	D	23	Glu	E	34
Phe	F	14	Gly	G	14	His	H	12	Ile	I	12
Lys	K	20	Leu	L	29	Met	M	5	Asn	N	14
Pro	P	18	Gln	Q	10	Arg	R	10	Ser	S	7
Thr	T	17	Val	V	20	Trp	W	4	Tyr	Y	7

Mol. wt (calc.) = 34 593   Residues = 303

```
  1    MRAWIFFLLC   LAGRALAAPQ   QEALPDETEV   VEETVAEVTE   VSVGANPVQV
 51    EVGEFDDGAE   ETEEEVVAEN   PCQNHHCKHG   KVCELDENNT   PMCVCQDPTS
101    CPAPIGEFEK   VCSNDNKTFD   SSCHFFATKC   TLEGTKKGHK   LHLDYIGPCK
151    YIPPCLDSEL   TEFPLRMRDW   LKNVLVTLYE   RDEDNNLLTE   KQKLRVKKIH
```

```
201  ENEKRLEAGD  HPVELLARDF  EKNYNMYIFP  VHWQFGQLDQ  HPIDGYLSHT
251  ELAPLRAPLI  PMEHCTTRFF  ETCDLDNDKY  IALDEWAGCF  GIKQKDIDKD
301  LVI
```

Structural and functional sites
Signal peptide: 1–17
Domain I: 23–70
Domain II: 71–155
Domain III: 185–221
Domain IV: 274–285
Ovomucoid/serine protease inhibitor repeat: 93–149
Potential N-linked glycosylation site: 116

Gene structure

The human gene is located on chromosome 5 at locus q31–q33 and contains ten exons. Although osteonectin is a single-gene product, an extracellular matrix glycoprotein (SC1) of brain tissue contains a COOH-terminal segment that is highly homologous to the protein encoded by exons 5–9 of osteonectin [3-6].

References
1 Romberg, R.W. et al (1985) Isolation and characterisation of native adult osteonectin. J. Biol. Chem. 260: 2728–2736.
2 Lankat-Buttgereit, B. et al (1988) Cloning and complete amino acid sequences of human and murine basement membrane protein BM–40 (SPARC-osteonectin). FEBS Lett. 236: 352–356.
3 McVey, J.H. et al (1988) Characterisation of the mouse SPARC/osteonectin gene. Intron/exon organisation and an unusual promoter region. J. Biol. Chem. 263: 11111–11116.
4 Swaroop, A. et al. (1988) Molecular analysis of the cDNA for human SPARC/osteonectin/BM–40: Sequence, expression, and localization of the gene to chromosome 5q31–q33. Genomics 2: 37–47.
5 Nomura, S. et al (1989) Evidence for positive and negative regulatory elements in the 5' flanking sequence of the mouse SPARC (osteonectin) gene. J. Biol. Chem. 264: 12201–12207.
6 Young, M.F. et al (1989) Osteonectin promoter. DNA sequence analysis and S1 endonuclease site potentially associated with transcriptional control in bone cells. J. Biol. Chem. 264: 450–456.
7 Johnston, I.G. et al (1990) Molecular cloning of SC1: A putative brain extracellular matrix glycoprotein showing partial similarity to osteonectin/BM40/SPARC. Neuron 2: 165–176.

Osteopontin

OPN, 2ar, secreted phosphoprotein 9SPPI), bone sialoprotein I (BSP I)

Osteopontin was initially characterized as an abundant bone sialoprotein. Subsequently it was found in a number of tissues, including placenta, distal tubules of the kidney, and the central nervous system. The molecule is expressed early in bone development, at high levels at sites of bone remodelling, and has been implicated in the process of osteogenesis. It binds to hydroxyapatite and appears to be associated with the attachment of osteoclasts, possibly via the integrin αVβ3. Osteopontin expression is stimulated by 1,25-dihydroxyvitamin D3, growth factors and tumour-promoting factors.

Molecular structure

Osteopontin is a single-chain polypeptide with a molecular weight of approximately 32 600. It is glycosylated (5–6 O-linked and one N-linked oligosaccharides in rat osteopontin), highly phosphorylated (12 phospho-Ser and one phospho-Thr in rat osteopontin), and sulphated. Secondary structure predictions suggest an open, extended, flexible structure. The molecule contains a stretch of negatively charged amino acids, principally aspartic acid, and a cell-binding RGD sequence [1–8].

Isolation
Osteopontin is isolated from bone by extraction with 4 M guanidine–HCl, demineralization with 0.5 M EDTA, size fractionation, and hydroxyapatite and ion-exchange chromatography [7].

Accession number
P10451

Primary structure

Ala	A	16	Cys	C	3	Asp	D	48	Glu	E	27
Phe	F	8	Gly	G	7	His	H	16	Ile	I	10
Lys	K	19	Leu	L	18	Met	M	5	Asn	N	12
Pro	P	15	Gln	Q	14	Arg	R	10	Ser	S	42
Thr	T	15	Val	V	19	Trp	W	2	Tyr	Y	8

Mol. wt (calc.) = 35 385   Residues = 314

```
1    MRIAVICFCL  LGITCAIPVK  QADSGSSEEK  QLYNKYPDAV  ATWLNPDPSQ
51   KQNLLAPQNA  VSSEETNDFK  QETLPSKSNE  SHDHMDDMDD  EDDDDHVDSQ
101  DSIDSNDSDD  VDDTDDSHQS  DESHHSDESD  ELVTDFPTDL  PATEVFTPVV
151  PTVDTYDGRG  DSVVYGLRSK  SKKFRRPDIQ  YPDATEDIT   SHMESEELNG
201  AYKAIPVAQD  LNAPSDWDSR  GKDSYETSQL  DDQSAETHSH  KQSRLYKRKA
251  NDESNEHSDV  IDSQELSKVS  REFHSHEFHS  HEDMLVVDPK  SKEEDKHLKF
301  RISHELDSAS  SEVN
```

Structural and functional sites
Signal peptide: 1–16
Potential N-linked glycosylation sites: 79, 106
RGD cell adhesion site: 159–161

Gene structure

The osteopontin gene is located on human chromosome 4 [6].

References

[1] Oldberg, A. et al (1986) Cloning and sequence analysis of rat bone sialoprotein (osteopontin) cDNA reveals an Arg–Gly–Asp cell binding sequence. Proc. Natl Acad. Sci. USA 83: 8819–8823.

[2] Fisher, L.W. et al (1987) Purification and partial characterisation of small proteoglycans I and II, bone sialoproteins I and II and osteonectin from the mineral compartment of developing human bone. J. Biol. Chem. 262: 9702–9708.

[3] Prince, C.W. et al (1987) Isolation, characterization and biosynthesis of a phosphorylated glycoprotein from rat bone. J. Biol. Chem. 262: 2900–2907.

[4] Yoon, K. et al (1987) Tissue specificity and developmental expression of rat osteopontin. Biochem. Biophys. Res. Commun. 148: 1129–1136.

[5] Kiefer, M.C. et al (1989) The cDNA and derived amino acid sequence for human osteopontin. Nucleic Acids Res. 17: 3306.

[6] Young, M.F. et al (1990) cDNA cloning, mRNA distribution and heterogeneity, chromosomal location, and RFLP analysis of human osteopontin (OPN). Genomics 7: 491–502.

[7] Zhang, Q. et al (1990) Characterisation of foetal porcine bone sialoproteins, secreted phosphoprotein I (SPPI, osteopontin), bone sialoprotein and a 23kDa glycoprotein. J. Biol. Chem. 265: 7583–7589.

[8] Sodek, J. et al (1992) Elucidating the functions of bone sialoprotein and osteopontin in bone formation. In: Chemistry and Biology of Mineralized Tissues, Slavkin, H. and Price, P., eds, Elsevier, Amsterdam, pp. 297–306.

Perlecan is a large proteoglycan and is a specific and integral component of all basement membranes. It endows basement membranes with fixed negative electrostatic charge and is responsible, in part, for the charge-selective ultrafiltration properties of this extracellular matrix. It interacts with other basement membrane components such as laminin and collagen type IV and serves as an attachment substrate for cells.

Molecular structure

Perlecan consists of a large protein core of molecular weight 470 000 to which three heparan sulphate glycosaminoglycan side-chains are attached. The size of the side-chains varies depending upon the source of the proteoglycan. Perlecan appears to consist of a series of globular domains separated by thin segments by electron microscopy after rotary shadowing. The heparan sulphate side-chains are located near one end of the protein core. The perlecan core protein can be divided into five domains. Domain I is the NH_2-terminal globular domain that includes three putative heparan sulphate glycosaminoglycan-attachment sites. Domain II contains four copies of internal repeats homologous to the ligand-binding domain of the LDL receptor. Domain II is separated from domain III by a short segment that shows homology to an immunoglobulin repeat. Domain III exhibits marked homology to the short arms of laminin chains. This domain includes four cysteine-rich subdomains containing eight-cysteine EGF repeats (CR-1 to CR-4) separated by three apparently globular subdomains (G1 to G3). These two types of subdomain exhibit homology to the laminin chain domains III/V and IV, respectively. Domain IV is composed of 21 consecutive immunoglobulin repeats similar to that separating domains II and III. The COOH-terminal domain V is similar to the COOH-terminal G domain of the laminin A chain. Domain V is composed of three homologous globular repeats (GR-1 to GR-3) separated by two double six-cysteine EGF repeats [1-3].

Heparan
sulphate
attachment

Isolation

Basement membrane heparan sulphate proteoglycan can be isolated after guanidine–HCl extraction by a combination of ion-exchange chromatography, size exclusion chromatography and caesium chloride density gradient centrifugation [1].

Accession number

X62515

Primary structure

Ala	A	340	Cys	C	188	Asp	D	171	Glu	E	240
Phe	F	107	Gly	G	443	His	H	166	Ile	I	144
Lys	K	69	Leu	L	353	Met	M	42	Asn	N	106
Pro	P	349	Gln	Q	256	Arg	R	264	Ser	S	407
Thr	T	271	Val	V	326	Trp	W	41	Tyr	Y	110

Mol. wt (calc.) = 468 863   Residues = 4393

```
1     MGWRAPGALL   LALLLHGRLL   AVTHGLRAYD   GLSLPEDIET   VTASQMRWTH
51    SYLSDDEDML   ADSISGDDLG   SGDLGSGDFQ   MVYFRALVNF   TRSIEYSPQL
101   EDAGSREFRE   VSEAVVDTLE   SEYLKIPGDQ   VVSVVFIKEL   DGWVFVELDV
151   GSEGNADGAQ   IQEMLLRVIS   SGSVASYVTS   PQGFQFRRLG   TVPQFPRACT
201   EAEFACHSYN   ECVALEYRCD   RRPDCRDMSD   ELNCEEPVLG   ISPTFSLLVE
251   TTSLPPRPET   TIMRQPPVTH   APQPLLPGSV   RPLPCGPQEA   ACRNGHCIPR
301   DYLCDGQEDC   EDGSDELDCG   PPPPCEPNEF   PCGNGHCALK   LWRCDGDFDC
351   EDRTDEANCP   TKRPEEVCGP   TQFRCVSTNM   CIPASFHCDE   ESDCPDRSDE
401   FGCMPPQVVT   PPRESIQASR   GQTVTFTCVA   IGVPAPFLIN   WRLNWGHIPS
451   QPRVTVTSEG   GRGTLIIRDV   KESDQGAYTC   EAMNARGMVF   GIPDGVLELV
501   PQRAGPCPDG   HFYLEHSAAC   LPCFCFGITS   VCQSTRRFRD   QIRLRFDQPD
551   DFKGVNVTMP   AQPGTPPLSS   TQLQIDPSLH   EFQLVDLSRR   FLVHDSFWAL
601   PEQFLGNKVD   SYGGSLRYNV   RYELARGMLE   PVQRPDVVLV   GAGYRLLSRG
651   HTPTQPGALN   QRQVQFSEEH   WVHESGRPVQ   RAELLQVLQS   LEAVLIQTVY
701   NTKMASVGLS   DIAMDTTVTH   ATSHGRAHSV   EECRCPIGYS   GLSCESCDAH
751   FTRVPGGPYL   GTCSGCSCNG   HASSCDPVYG   HCLNCQHNTE   GPQCKKCKAG
801   FFGDAMKATA   TSCRPCPCPY   IDASRRFSDT   CFLDTDGQAT   CDACAPGYTG
851   RRCESCAPGY   EGNPIQPGGK   CRPVNQEIVR   CDERGSMGTS   GEACRCKNNV
901   VGRLCNECAD   RSFHLSTRNP   DGCLKCFCMG   VSRHCTSSSW   SRAQLHGASE
951   EPGHFSLTNA   ASTHTTNEGI   FSPTPGELGF   SSFHRLLSGP   YFWSLPSRFL
1001  GDKVTSYGGE   LRFTVTQRSQ   PGSTPLHGQP   LVVLQGNNII   LEHHVAQEPS
1051  PGQPSTFIVP   FREQAWQRPD   GQPATREHLL   MALAGIDTLL   IRASYAQQPA
1101  ESRVSGISMD   VAVPEETGQD   PALEVEQCSC   PPGYRGPSCQ   DCDTGYTRTP
1151  SGLYLGTCER   CSCHGHSEAC   EPETGACQGC   QHHTEGPRCE   QCQPGYYGDA
1201  QRGTPQDCQL   CPCYGDPAAG   QAAHTCFLDT   DGHPTCDACS   PGHSGRHCER
1251  CAPGYYGNPS   QGQPCQRDSQ   VPGPIGCNCD   PQGSVSSQCD   AAGQCQCKAQ
1301  VEGLTCSHCR   PHHFHLSASN   PDGCLPCFCM   GITQQCASSA   YTRHLISTHF
1351  APGDFQGFAL   VNPQRNSRLT   GEFTVEPVPE   GAQLSFGNFA   QLGHESFYWQ
1401  LPETYQGDKV   AAYGGKLRYT   LSYTAGPQGS   PLSDPDVQIT   GNNIMLVASQ
1451  PALQGPERRS   YEIMFREEFW   RRPDGQPATR   EHLLMALADL   DELLIRATFS
1501  SVPLVASISA   VSLEVAQPGP   SNRPRALEVE   ECRCPPGYIG   LSCQDCAPGY
1551  TRTGSGLYLG   HCELCECNGH   SDLCHPETGA   CSQCQHNAAG   EFCELCAPGY
1601  YGDATAGTPE   DCQPCACPLT   NPENMFSRTC   ESLGAGGYRC   TACEPGYTGQ
1651  YCEQCGPGYV   GNPSVQGGQC   LPETNQAPLV   VEVHPARSIV   PQGGSHSLRC
1701  QVSGRGPHYF   YWSREDGRPV   PSGTQQRHQG   SELHFPSVQP   SDAGVYICTC
1751  RNLHRSNTSR   AELLVTEAPS   KPITVTVEEQ   RSQSVRPGAD   VTFICTAKSK
1801  SPAYTLVWTR   LHNGKLPTRA   MDFNGILTIR   NVQLSDAGTY   VCTGSNMFAM
1851  DQGTATLHVQ   ASGTLSAPVV   SIHPPQLTVQ   PGQLAEFRCS   ATGSPTPTLE
1901  WTGGPGGQLP   AKAQIHGGIL   RLPAVEPTDQ   AQYLCRAHSS   AGQQVARAVL
1951  HVHGGGGPRV   QVSPERTQVH   AGRTVRLYCR   AAGVPSATIT   WRKEGGSLPP
2001  QARSERTDIA   TLLIPAITTA   DAGFYLCVAT   SPAGTAQARI   QVVVLSASDA
2051  SQPPVKIESS   SPSVTEGQTL   DLNCVVAGSA   HAQVTWYRRG   GSLPHHTQVH
```

135

2101	GSRLRLPQVS	PADSGEYVCR	VENGSGPKEA	SITVSVLHGT	HSGPSYTPVP
2151	GSTRPIRIEP	SSSHVAEGQT	LDLNCVVPGQ	AHAQVTWHKR	GGSLPARHQT
2201	HGSLLRLHQV	TPADSGEYVC	HVVGTSGPLE	ASVLVTIEAS	VIPGPIPPVR
2251	IESSSSTVAE	GQTLDLSCVV	AGQAHAQVTW	YKRGGSLPAR	HQVRGSRLYI
2301	FQASPADAGQ	YVCRASNGME	ASITVTVTGT	QGANLAYPAG	STQPIRIEPS
2351	SSQVAEGQTL	DLNCVVPGQS	HAQVTWHKRG	GSLPVRHQTH	GSLLRLYQAS
2401	PADSGEYVCR	VLGSSVPLEA	SVLVTIEPAG	SVPALGVTPT	VRIESSSSQV
2451	AEGQTLDLNC	LVAGQAHAQV	TWHKRGGSLP	ARHQVHGSRL	RLLQVTPADS
2501	GEYVCRVVGS	SGTQEASVLV	TIQQRLSGSH	SQGVAYPVRI	ESSSASLANG
2551	HTLDLNCLVA	SQAPHTITWY	KRGGSLPSRH	QIVGSRLRIP	QVTPADSGEY
2601	VCHVSNGAGS	RETSLIVTIQ	GSGSSHVPRV	SPPIRIESSS	PTVVEGQTLD
2651	LNCVVARQPQ	AIITWYKRGG	SLPSRHQTHG	SHLRLHQMSV	ADSGEYVCRA
2701	NNNIDALEAS	IVISVSPSAG	SPSAPGSSMP	IRIESSSSHV	AEGETLDLNC
2751	VVPGQAHAQV	TWHKRGGSLP	SYHQTRGSRL	RLHHVSPADS	GEYVCRVMGS
2801	SGPLEASVLV	TIEASGSSAV	HVPAPGGAPP	IRIEPSSSRV	AEGQTLDLKC
2851	VVPGQAHAQV	TWHKRGGNLP	ARHQVHGPLL	RLNQVSPADS	GEYSCQVTGS
2901	SGTLEASVLV	TIEPSSPGPI	PAPGLAQPIY	IEASSSHVTE	GQTLDLNCVV
2951	PGQAHAQVTW	YKRGGSLPAR	HQTHGSQLRL	HHVSPADSGE	YVCRAAGGPG
3001	PEQEASFTVT	VPPSEGSSYR	LRSPVISIDP	PSSTVQQGQD	ASFKCLIHDG
3051	AAPISLEWKT	RNQELEDNVH	ISPNGSIITI	VGTRPSNHGT	YRCVASNAYG
3101	VAQSVVNLSV	HGPPTVSVLP	EGPVWVKVGK	AVTLECVSAG	EPRSSARWTR
3151	ISSTPAKLEQ	RTYGLMDSHT	VLQISSAKPS	DAGTYVCLAQ	NALGTAQKQV
3201	EVIVDTGAMA	PGAPQVQAEE	AELTVEAGHT	ATLRCSATGS	PARTIHWSKL
3251	RSPLPWQHRL	EGDTLIIPRV	AQQDSGQYIC	NATSPAGHAE	ATIILHVESP
3301	PYATTVPEHA	SVQAGETVQL	QCLAHGTPPL	TFQWSRVGSS	LPGRATARNE
3351	LLHFERAAPE	DSGRYRCRVT	NKVGSAEAFA	QLLVQGPPGS	LPATSIPAGS
3401	TPTVQVTPQL	ETKSIGASVE	FHCAVPSDRG	TQLRWFKEGG	QLPPGHSVQD
3451	GVLRIQNLDQ	SCQGTYICQA	HGPWGKAQAS	AQLVIQALPS	VLINIRTSVQ
3501	TVVVGHAVEF	ECLALGDPKP	QVTWSKVGGH	LRPGIVQSGG	VVRIAHVELA
3551	DAGQYRCTAT	NAAGTTQSHV	LLLVQALPQI	SMPQEVRVPA	GSAAVFPCIA
3601	SGYPTPDISW	SKLDGSLPPD	SRLENNMLML	PSVQPQDAGT	YVCTATNRQG
3651	KVKAFAHLQV	PERVVPYFTQ	TPYSFLPLPT	IKDAYRKFEI	KITFRPDSAD
3701	GMLLYNGQKR	VPGSPTNLAN	RQPDFISFGL	VGGRPEFRFD	AGSGMATIRH
3751	PTPLALGHFH	TVTLLRSLTQ	GSLIVGDLAP	VNGTSQGKFQ	GLDLNEELYL
3801	GGYPDYGAIP	KAGLSSGFIG	CVRELRIQGE	EIVFHDLNLT	AHGISHCPTC
3851	RDRPCQNGGQ	CHDSESSSYV	CVCPAGFTGS	RCEHSQALHC	HPEACGPDAT
3901	CVNRPDGRGY	TCRCHLGRSG	LRCEEGVTVT	TPSLSGAGSY	LALPALTNTH
3951	HELRLDVEFK	PLAPDGVLLF	SGGKSGPVED	FVSLAMVGGH	LEFRYELGSG
4001	LAVLRTAEPL	ALGRWHRVSA	ERLNKDGSLR	VNGGRPVLRS	SPGKSQGLNL
4051	HTLLYLGGVE	PSVPLSPATN	MSAHFRGCVG	EVSVNGKRLD	LTYSFLGSQG
4101	IGQCYDSSPC	ERQPCQHGAT	CMPAGEYEFQ	CLCRDGIKGD	LCEHEENPCQ
4151	LREPCLHGGT	CQGTRCLCLP	GFSGPRCQQG	SGHGIAESDW	HLEGSGGNDA
4201	PGQYGAYFHD	DGFLAFPGHV	FSRSLPEVPE	TIELEVRTST	ASGLLLWQGV
4251	EVGEAGQGKD	FISLGLQDGH	LVFRYQLGSG	EARLVSEDPI	NDGEWHRVTA
4301	LREGRRGSIQ	VDGEELVSGR	SPGPNVAVNA	KGSIYIGGAP	DVATLTGGRF
4351	SSGITGCVKN	LVLHSARPGA	PPPQPLDLQH	RAQAGANTRP	CPS

Structural and functional sites

Signal peptide: 1–21
LDL receptor repeats: 194–234, 281–319, 320–359, 360–403

Immunoglobulin repeats: 404–506, 1679–1773, 1774–1867, 1868–1957, 1958–2052, 2053–2153, 2154–2246, 2247–2342, 2343–2438,2439–2535, 2536–2631, 2632–2728, 2739–2828, 2829–2926, 2927–3023, 3024–3114, 3115–3213, 3214–3300, 3301–3401, 3402–3490, 3491–3576, 3577–3676
EGF (8C) repeats: 507–532 (partial), 733–765 (partial), 766–815, 816–873, 874–925 (partial), 926–935 (partial), 1127–1160 (partial), 1161–1210, 1211–1276, 1277–1326, 1327–1336 (partial), 1532–1564 (partial), 1565–1614, 1615–1672
EGF (6C) repeats: 3850–3889, 3890–3930, 4110–4148, 4149–4184
G repeats: 3689–3849, 3952–4109, 4233–4393
Putative glycosaminoglycan-attachment sites: 65, 71, 75
Potential N-linked glycosylation sites: 89, 556, 1757, 2123, 3074, 3107, 3281, 3782, 3838, 4070

Gene structure

The perlecan gene has been localized to human chromosome 1p35–36 [2].

References

[1] Paulsson, M. et al (1987) Structure of low density heparan sulphate proteoglycan isolated from a mouse tumor basement membrane. J. Mol. Biol. 197: 297–313.
[2] Kallunki, P. et al (1991) Cloning of human heparan sulphate proteoglycan core protein, assignment of the gene (HSPG1) to 1p36.1–p35 and identification of a BamHI restriction fragment length polymorphism. Genomics 11: 389–396.
[3] Kallunki, P. and Tryggvason, K. (1992) Human basement membrane heparan sulphate proteoglycan core protein: A 467-kD protein containing multiple domains resembling elements of the low density lipoprotein receptor, laminin, neural cell adhesion molecules and epidermal growth factor. J. Cell Biol. 116: 559–571.

Phosphoryn

dentin matrix acidic phosphoprotein, AG1

Phosphoryn encompasses a group of highly phosphorylated proteins which are present in dentine extracellular matrix. There appear to be species-specific differences in phosphoryn composition since in bovine phosphoryn there is only one type of molecule while in the rat incisor version two or three types of molecule are found. These differences may be explained by age-dependent degradation *in situ*. Phosphoryn has been shown to markedly affect *in vitro* crystallization of hydroxyapatite, and since the molecules are secreted directly at the mineral front they are suggested to play a major role in dentinogenesis.

Molecular structure

In solution at neutral pH and physiological ionic strength, the rat incisor phosphoryn molecules are folded as loose globular structures, with diameters of about 20 nm. These molecules bind to both collagen monomers and collagen fibres. Each phosphoryn molecule is a single polypeptide chain which is highly phosphorylated and rich in acidic amino acids. Aspartic acid and phosphoserine comprise over 80% of all residues in some molecules isolated, but AG1 contains approximately equal amounts of aspartic acid and glutamic acid residues and has a composition intermediate between the glutamic acid-rich phosphoproteins of bone and the aspartic acid-rich proteins of dentine. Approximately half of the glutamic acid residues appear as consecutive EE sequences, while about 55 of the 107 serine residues can be readily phosphorylated. Considerable discrepancies exist in the literature concerning molecular size, due mainly to high charge density and consequent anomalous biochemical behaviour [1-6].

Isolation
EDTA-soluble phosphoryn can be purified from dentine by sequential calcium chloride precipitation, gel filtration in SDS buffer, anion-exchange chromatography and gel filtration in 4 M guanidine–HCl [7].

Accession number
L11354

Primary structure

Ala	A	20	Cys	C	2	Asp	D	62	Glu	E	72
Phe	F	7	Gly	G	36	His	H	8	Ile	I	5
Lys	K	11	Leu	L	15	Met	M	5	Asn	N	17
Pro	P	19	Gln	Q	33	Arg	R	29	Ser	S	107
Thr	T	26	Val	V	8	Trp	W	2	Tyr	Y	5

Mol. wt (calc.) = 53 000   Residues = 489

```
  1    MKTVILLTFL   WGLSCALPVA   RYQNTESESS   EGRTGNLAQS   PFPFMANSDH
 51    TDSSESGEEL   GSDRSQYRFA   GGLSKSAGMD   ADKEEDEDDS   GDDTFGDEDN
101    GPGPEERQWG   GPSRLDSDED   SADTTQSSED   STSQENSAQD   TPSDSKDHHS
151    DEADSRPEAG   DSTQDSESEG   YRVGGGSEGE   SSHGDGSEFD   DEGMQSDDPG
201    STRSDRGHTR   MSSADISSGE   SKGDHEPTST   QDSDDSQDVE   FSSRKSFRRS
251    RVSEEDDRGE   LADSNSRGTQ   SVSTEDFRSK   EESDSETQGD   TAETQSQEDS
301    PEGQDPSSES   SEEAGEPSQE   SSSGSQEGVA   SESRGDNPDN   TSQTGDQRDS
```

```
351   ESSEEDRLNT   FSSSESQSTE   EQGDSESNES   LSLSEESQES   AQDEDSSSQE
401   GLQSQSASRE   SRSQESQSEG   RSRSEENRDS   DSQDSSRSKE   ESNSTGSTSS
451   SEEDNIFKNI   EADNRKLIVD   AYHNKPIGDE   DDNDCQDGY
```

Structural and functional sites

Signal peptide: 1–16 (putative)
Potential N-linked glycosylation site: 340
Acidic patches: 84–89, 254–257
RGD putative cell adhesion site: 334–336

Gene structure

Not known.

References

[1] DiMuzio, M.T. and Veis, A. (1978) Phosphoryns. Major noncollagenous proteins of rat incisor dentine. Calcified Tissue Res. 25: 169–178.

[2] Butler, W.T. et al (1983) Multiple forms of rat dentin phosphoproteins. Arch. Biochem. Biophys. 225: 178–186.

[3] Nakamura, O. et al (1985) Immunohistochemical studies with a monoclonal antibody on the distribution of phosphoryn in predentin and dentin. Calcified Tissue Int. 37: 491–500.

[4] Ibaraki, K. et al (1991) An analysis of the biochemical and biosynthetic properties of dentin phosphoprotein. Matrix 11: 115–124.

[5] Sabsay, B. et al (1991) Domain structure and sequence distribution in dentin phosphophoryn. Biochem. J. 276: 699–707.

[6] George, A. et al (1993) Characterisation of a novel dentin matrix acidic phosphoprotein. Implications for induction of biomineralization. J. Biol. Chem. (in press).

[7] Stetler-Stevenson, W.G. and Veis, A. (1983) Bovine dentin phosphoryn: Composition and molecular weight. Biochemistry 22: 4326–4335.

Tenascin

hexabrachion, myotendinous antigen, J1, cytotactin

Tenascin is a polymorphic, high molecular weight extracellular matrix glycoprotein. It is transiently expressed in many developing organs and reappears in the stroma of many tumours. This highly regulated expression suggests a possible function in cell–matrix adhesion and cell migration, and in modulation of growth and differentiation during morphogenesis. Current evidence suggests that tenascin exhibits an anti-adhesive activity.

Molecular structure

Tenascin is composed of six similar subunits joined at their NH_2-termini by disulphide bonds. The difference between the subunits is determined by alternative splicing. A striking feature of the sequence are its multiple repeats. Three different types of repeat have been identified; six-cysteine EGF repeats, fibronectin type III repeats and a region homologous to the β and γ chains of fibrinogen. The junctional region is proposed to be formed by a triple coiled-coil of three subunits mediated by four heptad repeats of hydrophobic amino acids. The central globular domain is formed by a contribution from all six subunits and is the site at which subunits are fixed in place by disulphide cross-linking. The putative RGD cell adhesion sequence at position 876–878 is not conserved between species and is probably not functional [1-4].

Isolation

Tenascin can be isolated from conditioned medium after removal of fibronectin by gelatin–Sepharose affinity chromatography followed by hydroxyapatite chromatography. Final purification can be achieved by two sequential precipitations with 6% and 12.8% polyethylene glycol [5].

Accession number

P24821

Primary structure

Ala	A	142	Cys	C	99	Asp	D	134	Glu	E	178
Phe	F	67	Gly	G	194	His	H	39	Ile	I	98
Lys	K	81	Leu	L	167	Met	M	23	Asn	N	94
Pro	P	112	Gln	Q	76	Arg	R	119	Ser	S	139
Thr	T	190	Val	V	154	Trp	W	28	Tyr	Y	65

Mol. wt (calc.) = 240 449   Residues = 2199

```
1    MGAMTQLLAG  VFLAFLALAT  EGGVLKKVIR  HKRQSGVNAT  LPEENQPVVF
51   NHVYNIKLPV  GSQCSVDLES  ASGEKDLAPP  SEPSESFQEH  TVDGENQIVF
101  THRINIPRRA  CGCAAAPDVK  ELLSRLEELE  NLVSSLREQC  TAGAGCCLQP
151  ATGRLDTRPF  CSGRGNFSTE  GCGCVCEPGW  KGPNCSEPEC  PGNCHLRGRC
201  IDGQCICDDG  FTGEDCSQLA  CPSDCNDQGK  CVNGVCICFE  GYAADCSREI
251  CPVPCSEEHG  TCVDGLCVCH  DGFAGDDCNK  PLCLNNCYNR  GRCVENECVC
301  DEGFTGEDCS  ELICPNDCFD  RGRCINGTCY  CEEGFTGEDC  GKPTCPHACH
351  TQGRCEEGQC  VCDEGFAGVD  CSEKRCPADC  HNRGRCVDGR  CECDDGFTGA
```

```
401   DCGELKCPNG   CSGHGRCVNG   QCVCDEGYTG   EDCSQLRCPN   DCHSRGRCVE
451   GKCVCEQGFK   GYDCSDMSCP   NDCHQHGRCV   NGMCVCDDGY   TGEDCRDRQC
501   PRDCSNRGLC   VDGQCVCEDG   FTGPDCAELS   CPNDCHGQGR   CVNGQCVCHE
551   GFMGKDCKEQ   RCPSDCHGQG   RCVDGQCICH   EGFTGLDCGQ   HSCPSDCNNL
601   GQCVSGRCIC   NEGYSGEDCS   EVSPPKDLVV   TEVTEETVNL   AWDNEMRVTE
651   YLVVYTPTHE   GGLEMQFRVP   GDQTSTIIRE   LEPGVEYFIR   VFAILENKKS
701   IPVSARVATY   LPAPEGLKFK   SIKETSVEVE   WDPLDIAFET   WEIIFRNMNK
751   EDEGEITKSL   RRPETSYRQT   GLAPGQEYEI   SLHIVKNNTR   GPGLKRVTTT
801   RLDAPSQIEV   KDVTDTTALI   TWFKPLAEID   GIELTYGIKD   VPGDRTTIDL
851   TEDENQYSIG   NLKPDTEYEV   SLISRRGDMS   SNPAKETFTT   GLDAPRNLRR
901   VSQTDNSITL   EWRNGKAAID   SYRIKYAPIS   GGDHAEVDVP   KSQQATTKTT
951   LTGLRPGTEY   GIGVSAVKED   KESNPATINA   ATELDTPKDL   QVSETAETSL
1001  TLLWKTPLAK   FDRYRLNYSL   PTGQWVGVQL   PRNTTSYVLR   GLEPGQEYNV
1051  LLTAEKGRHK   SKPARVKAST   EQAPELENLT   VTEVGWDGLR   LNWTAADQAY
1101  EHFIIQVQEA   NKVEAARNLT   VPGSLRAVDI   PGLKAATPYT   VSIYGVIQGY
1151  RTPVLSAEAS   TGETPNLGEV   VVAEVGWDAL   KLNWTAPEGA   YEYFFIQVQE
1201  ADTVEAAQNL   TVPGGLRSTD   LPGLKAATHY   TITIRGVTQD   FSTTPLSVEV
1251  LTEEVPDMGN   LTVTEVSWDA   LRLNWTTPDG   TYDQFTIQVQ   EADQVEEAHN
1301  LTVPGSLRSM   EIPGLRAGTP   YTVTLHGEVR   GHSTRPLAVE   VVTEDLPQLG
1351  DLAVSEVGWD   GLRLNWTAAD   NAYEHFVIQV   QEVNKVEAAQ   NLTLPGSLRA
1401  VDIPGLEAAT   PYRVSIYGVI   RGYRTPVLSA   EASTAKEPEI   GNLNVSDITP
1451  ESFNLSWMAT   DGIFETFTIE   IIDSNRLLET   VEYNISGAER   TAHISGLPPS
1501  TDFIVYLSGL   APSIRTKTIS   ATATTEALPL   LENLTISDIN   PYGFTVSWMA
1551  SENAFDSFLV   TVVDSGKLLD   PQEFTLSGTQ   RKLELRGLIT   GIGYEVMVSG
1601  FTQGHQTKPL   RAEIVTEAEP   EVDNLLVSDA   TPDGFRLSWT   ADEGVFDNFV
1651  LKIRDTKKQS   EPLEITLLAP   ERTRDITGLR   EATEYEIELY   GISKGRRSQT
1701  VSAIATTAMG   SPKEVIFSDI   TENSATVSWR   APTAQVESFR   ITYVPITGGT
1751  PSMVTVDGTK   TQTRLVKLIP   GVEYLVSIIA   MKGFEESEPV   SGSFTTALDG
1801  PSGLVTANIT   DSEALARWQP   AIATVDSYVI   SYTGEKVPEI   TRTVSGNTVE
1851  YALTDLEPAT   EYTLRIFAEK   GPQKSSTITA   KFTTDLDSPR   DLTATEVQSE
1901  TALLTWRPPR   ASVTGYLLVY   ESVDGTVKEV   IVGPDTTSYS   LADLSPSTHY
1951  TAKIQALNGP   LRSNMIQTIF   TTIGLLYPFP   KDCSQAMLNG   DTTSGLYTIY
2001  LNGDKAQALE   VFCDMTSDGG   GWIVFLRRKN   GRENFYQNWK   AYAAGFGDRR
2051  EEFWLGLDNL   NKITAQGQYE   LRVDLRDHGE   TAFAVYDKFS   VGDAKTRYKL
2101  KVEGYSGTAG   DSMAYHNGRS   FSTFDKDTDS   AITNCALSTR   GFWYRNCHRV
2151  NLMGRYGDNN   HSQGVNWFHW   KGHEHSIQFA   EMKLRPSNFR   NLEGRRKRA
```

Structural and functional sites

Signal peptide: 1–22

EGF (6C) repeats: 174–185, 186–216, 217–246, 247–278, 279–309, 310–340, 341–371, 372–402, 403–433, 434–464, 465–495, 496–526, 527–557, 558–588, 589–619

Fibronectin type III repeats: 620–709, 710–800, 801–890, 891–982, 983–1070, 1071–1161, 1162–1252, 1253–1343, 1344–1434, 1435–1525, 1526–1616, 1617–1707, 1708–1796, 1797–1884, 1885–1972

Alternatively spliced repeats: 983–1070, 1075–1252, 1253–1343, 1344–1434, 1435–1525, 1526–1616, 1617–1707

Potential N-linked glycosylation sites: 38, 166, 184, 326, 787, 1017, 1033, 1078, 1092, 1118, 1183, 1209, 1260, 1274, 1300, 1365, 1391, 1444, 1454, 1484, 1533, 1808, 2160

RGD cell adhesion site: 876–878 (not conserved in murine sequence)

Gene structure

The tenascin gene is located on human chromosome 9 at locus q32–34; however, Southern blot analyses indicate the presence of multiple related genes. The coding region of the tenascin gene spans approximately 80 kb and consists of 27 exons. The type III repeats are encoded by one or two exons. All alternatively spliced type III repeats are encoded by single exons. The fibrinogen-like domain is encoded by five exons and the EGF repeats by single exons.[6,7].

References
[1] Erickson, H.P. and Bourdon, M.A. (1989) Tenascin. An extracellular matrix protein prominent in specialised embryonic tissues and tumours. Annu. Rev. Cell Biol. 5: 71–92.

[2] Gulcher, J.R. et al (1989) Structure of the human hexabrachion (tenascin) gene. Proc. Natl Acad. Sci. USA 88: 9438–9442.

[3] Chiquet-Ehrismann, R. (1990) What distinguishes tenascin from fibronectin? FASEB J. 4: 2598–2604.

[4] Chiquet-Ehrismann, R. (1991) Anti-adhesive molecules of the extracellular matrix. Current Opinion Cell Biol. 3: 800–804.

[5] Saginata, M. et al (1992) A simple procedure for tenascin purification. Eur. J. Biochem. 205: 545–549.

[6] Gulcher, J.R. et al (1989) An alternative spliced region of the human hexabrachion contains a novel repeat of potential N-glycosylation sites. Proc. Natl Acad. Sci. USA 86: 1588–1592.

[7] Siri, A. et al (1991) Human tenascin, primary structure, pre-mRNA splicing patterns and localisation of the epitope recognised by 2 monoclonal antibodies. Nucleic Acids Res. 19: 525–531.

Thrombospondin

There are at least four different thrombospondin genes, but currently only thrombospondin 1 has been characterized in detail. This is the most abundant protein component of platelet α granules and is rapidly secreted upon platelet activation at sites of injury and thrombosis. Thrombospondin 1 is synthesized and secreted by a variety of cell types, including fibroblasts and smooth muscle cells and has been implicated in the regulation of cell migration and proliferation during development, wound healing, angiogenesis and tumorigenesis. Its adhesive interactions with cells are complicated and are mediated by a variety of receptors including integrins, CD36, proteoglycans and sulphatides. In addition, thrombospondin binds fibrinogen, fibronectin, laminin and collagens.

Molecular structure

Thrombospondin 1 is a trimer made up of three identical subunits joined by disulphide bonds. In platelets, the subunits have a molecular weight of approximately 180 000. In line with other extracellular adhesive glycoproteins, the subunits are made up of discrete structural domains. There are three repeating motifs; the three type 1 motifs are homologous to the complement components C6–C9 and properdin and contain the major cell-binding site, VTCG, which probably interacts with the counter-receptor CD36. There are three type 2 or six-cysteine EGF repeats, and eight cysteine-containing type 3 repeats which include either one or two copies of an EF-hand-type loop and which probably constitute the major calcium-binding site in the molecule. In addition to the three repeating motifs, there are several distinct segments to the molecule; these include a heparin-binding domain which participates in thrombospondin-mediated disruption of focal contacts and regulation of proliferation, a pair of cysteine residues that cross-link the three subunits, a 70 amino acid segment that is homologous to the N-propeptide of type I collagen, and a COOH-terminal domain that interacts with platelets and other cells. Preliminary reports suggest that alternative splicing can produce two variant chains of molecular weight 140 000 and 50 000 [1-7].

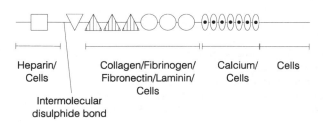

Heparin/Cells

Collagen/Fibrinogen/Fibronectin/Laminin/Cells

Calcium/Cells

Cells

Intermolecular disulphide bond

Isolation

Thrombospondin can be purified from the supernatant of activated platelets by passage through a gelatin–Sepharose column and retention on a heparin–Sepharose column. Thrombospondin can be further purified by size exclusion on Bio-Gel A-0.5m [8].

Accession number

P07996

Primary structure

Sequence conflicts: 84   T to A
                    523 T to A

Ala	A	51	Cys	C	70	Asp	D	131	Glu	E	51
Phe	F	35	Gly	G	105	His	H	27	Ile	I	45
Lys	K	55	Leu	L	64	Met	M	14	Asn	N	81
Pro	P	70	Gln	Q	50	Arg	R	60	Ser	S	74
Thr	T	62	Val	V	72	Trp	W	22	Tyr	Y	31

Mol. wt (calc.) = 129 270   Residues = 1170

```
1     MGLAWGLGVL  FLMHVCGTNR  IPESGGDNSV  FDIFELTGAA  RKGSGRRLVK
51    GPDPSSPAFR  IEDANLIPPV  PDDKFQDLVD  AVRTEKGFLL  LASLRQMKKT
101   RGTLLALERK  DHSGQVFSVV  SNGKAGTLDL  SLTVQGKQHV  VSVEEALLAT
151   GQWKSITLFV  QEDRAQLYID  CEKMENAELD  VPIQSVFTRD  LASIARLRIA
201   KGGVNDNFQG  VLQNVRFVFG  TTPEDILRNK  GCSSSTSVLL  TLDNNVVNGS
251   SPAIRTNYIG  HKTKDLQAIC  GISCDELSSM  VLELRGLRTI  VTTLQDSIRK
301   VTEENKELAN  ELRRPPLCYH  NGVQYRNNEE  WTVDSCTECH  CQNSVTICKK
351   VSCPIMPCSN  ATVPDGECCP  RCWPSDSADD  GWSPWSEWTS  CSTSCGNGIQ
401   QRGRSCDSLN  NRCEGSSVQT  RTCHIQECDK  RFKQDGGWSH  WSPWSSCSVT
451   CGDGVITRIR  LCNSPSPQMN  GKPCEGEARE  TKACKKDACP  INGGWGPWSP
501   WDICSVTCGG  GVQKRSRLCN  NPTPQFGGKD  CVGDVTENQI  CNKQDCPIDG
551   CLSNPCFAGV  KCTSYPDGSW  KCGACPPGYS  GNGIQCTDVD  ECKEVPDACF
601   NHNGEHRCEN  TDPGYNCLPC  PPRFTGSQPF  GQGVEHATAN  KQVCKPRNPC
651   TDGTHDCNKN  AKCNYLGHYS  DPMYRCECKP  GYAGNGIICG  EDTDLDGWPN
701   ENLVCVANAT  YHCKKDNCPN  LPNSGQEDYD  KDGIGDACDD  DDDNDKIPDD
751   RDNCPFHYNP  AQYDYDRDDV  GDRCDNCPYN  HNPDQADTDN  NGEGDACAAD
801   IDGDGILNER  DNCQYVYNVD  QRDTDMDGVG  DQCDNCPLEH  NPDQLDSDSD
851   RIGDTCDNNQ  DIDEDGHQNN  LDNCPYVPNA  NQADHDKDGK  GDACDHDDDN
901   DGIPDDKDNC  RLVPNPDQKD  SDGDGRGDAC  KDDFDHDSVP  DIDDICPENV
951   DISETDFRRF  QMIPLDPKGT  SQNDPNWVVR  HQGKELVQTV  NCDPGLAVGY
1001  DEFNAVDFSG  TFFINTERDD  DYAGFVFGYQ  SSSRFYVVMW  KQVTQSYWDT
1051  NPTRAQGYSG  LSVKVVNSTT  GPGEHLRNAL  WHTGNTPGQV  RTLWHDPRHI
1101  GWKDFTAYRW  RLSHRPKTGF  IRVVMYEGKK  IMADSGPIYD  KTYAGGRLGL
1151  FVFSQEMVFF  SDLKYECRDP
```

Structural and functional sites

Signal peptide: 1–18
Procollagen homology region: 303–372
Properdin repeats: 379–434, 435–491, 492–548
EGF (6C) repeats: 549–587, 588–645, 646–689
Thrombospondin type 3 repeats: 723–758, 759–781, 782–817, 818–840, 841–878, 879–914, 915–950
Potential N-linked glycosylation sites: 248, 360, 708, 1067
Heparin-binding sites: 41–47, 99–102
RGD cell adhesion site: 926–928
VTCG cell adhesion sites: 449–452, 506–509
Platelet adhesion sites: 1034–1063, 1084–1138
Alternatively spliced domain: 19–292

Gene structure

Thrombospondin 1 is located on human chromosome 15q15, thrombospondin 2 on chromosome 6q27, and thrombospondin 3 on chromosome 1q21–24 [9].

References

[1] Lawler, J. and Hynes, R.O. (1986) The structure of human thrombospondin, an adhesive glycoprotein with multiple calcium binding sites and homologies with several different proteins. J. Cell Biol. 103: 1635–1648.

[2] Paul, L. et al (1989) Identification of an alternatively spliced product of the human thrombospondin gene. J. Cell Biol. 109: 200a.

[3] Bornstein, P. (1992) Thrombospondins: Structure and regulation of expression. FASEB J. 6: 3290–3299.

[4] Kosfeld, M.D. and Frazier, W.A. (1992) Identification of active peptide sequences in the carboxyl-terminal cell binding domain of human thrombospondin–1. J. Biol. Chem. 267: 16230–16236.

[5] Laherty, C.D. et al (1992) Characterisation of mouse thrombospondin-2 sequence and expression during cell growth and development. J. Biol. Chem. 267: 3274–3281.

[6] Vos, H.L. et al (1992) Thrombospondin 3 (Thbs3), a new member of the thrombospondin gene family. J. Biol. Chem. 267: 12192–12196.

[7] Adams, J.C. and Lawler, J. (1993) The thrombospondin family. Current Opinion Cell Biol. (in press).

[8] Santoro, S.A. and Frazier, W.D. (1987) Isolation and characterisation of thrombospondin. Methods Enzymol. 144: 438–446.

[9] Wolf, F.W. et al (1990) Structure and chromosomal localisation of the human thrombospondin gene. Genomics 6: 685–691.

Versican

Versican is a large chondroitin sulphate proteoglycan secreted by fibroblasts. Versican contains domains that are highly homologous to aggrecan in having a hyaluronan-binding domain at its NH2-terminal end and two EGF repeats, a lectin repeat and a complement control protein repeat at its COOH-terminal end. Versican may play a role in intracellular signalling, cell recognition and connecting extracellular matrix components and cell-surface glycoproteins.

Molecular structure

The versican core protein is highly negatively charged and has a calculated pI of 4.2. The NH2-terminal domain is similar to the three-loop structure of link protein and to the link-like sequences of aggrecan. On the COOH-side of the hyaluronan-binding domain is a 200 amino acid domain containing two cysteines and a cluster of glutamic acid residues that may be important in the interaction of versican with hydroxyapatite in bone. The COOH-terminal domain of versican contains two EGF repeats in tandem and sequences homologous to the complement control protein repeats of murine factor H and human C4-binding protein [1–4].

Isolation

Versican is present in the culture medium of human IMR–90 lung fibroblasts. It has been purified from this source by ammonium sulphate precipitation and DEAE–cellulose chromatography [3].

Accession number

P13611

Primary structure

Sequence conflict: 1722–1726 IKAEA to EFREV

Ala	A	146	Cys	C	35	Asp	D	141	Glu	E	258
Phe	F	95	Gly	G	133	His	H	57	Ile	I	114
Lys	K	96	Leu	L	139	Met	M	37	Asn	N	73
Pro	P	155	Gln	Q	94	Arg	R	79	Ser	S	253
Thr	T	259	Val	V	159	Trp	W	24	Tyr	Y	62

Mol. wt (calc.) = 264 759   Residues = 2409

```
1     MFINIKSILW   MCSTLIVTHA   LHKVKVGKSP   PVRGSLSGKV   SLPCHFSTMP
51    TLPPSYNTSE   FLRIKWSKIE   VDKNGKDLKE   TTVLVAQNGN   IKIGQDYKGR
101   VSVPTHPEAV   GDASLTVVKL   LASDAGLYRC   DVMYGIEDTQ   DTVSLTVDGV
151   VFHYRAATSR   YTLNFEAAQK   ACLDVGAVIA   TPEQLFAAYE   DGFEQCDAGW
201   LADQTVRYPI   RAPRVGCYGD   KMGKAGVRTY   GFRSPQETYD   VYCYVDHLDG
251   DVFHLTVPSK   FTFEEAAKEC   ENQDARLATV   GELQAAWRNG   FDQCDYGWLS
301   DASVRHPVTV   ARAQCGGGLL   GVRTLYRFEN   QTGFPPPDSR   FDAYCFKRRM
351   SDLSVIGHPI   DSESKEDEPC   SEETDPVHDL   MAEILPEFPD   IIEIDLYHSE
401   ENEEEEEECA   NATDVTTTPS   VQYINGKHLV   TTVPKDPEAA   EARRGQFESV
451   APSQNFSDSS   ESDTHPFVIA   KTELSTAVQP   NESTETTESL   EVTWKPETYP
```

```
 501   ETSEHFSGGE   PDVFPTVPFH   EEFESGTAKK   GAESVTERDT   EVGHQAHEHT
 551   EPVSLFPEES   SGEIAIDQES   QKIAFARATE   VTFGEEVEKS   TSVTYTPTIV
 601   PSSASAYVSE   EEAVTLIGNP   WPDDLLSTKE   SWVEATPRQV   VELSGSSSIP
 651   ITEGSGEAEE   DEDTMFTMVT   DLSQRNTTDT   LITLDTSRII   TESFFEVPAT
 701   TIYPVSEQPS   AKVVPTKFVS   ETDTSEWISS   TTVEEKKRKE   EEGTTGTAST
 751   FEVYSSTQRS   DQLILPFELE   SPNVATSSDS   GTRKSFMSLT   TPTQSEREMT
 801   DSTPVFTETN   TLENLGAQTT   EHSSIHQPGV   QEGLTTLPRS   PASVFMEQGS
 851   GEAAADPETT   TVSSFSLNVE   YAIQAEKEVA   GTLSPHVETT   FSTEPTGLVL
 901   STVMDRVVAE   NITQTSREIV   ISERLGEPNY   GAEIRGFSTG   FPLEEDFSGD
 951   FREYSTVSHP   IAKEETVMME   GSGDAAFRDT   QTSPSTVPTS   VHISHISDSE
1001   GPSSTMVSTS   AFPWEEFTSS   AEGSGEQLVT   VSSSVVPVLP   SAVQKFSGTA
1051   SSIIDEGLGE   VGTVNEIDRR   STILPTAEVE   GTKAPVEKEE   VKVSGTVSTN
1101   FPQTIEPAKL   WSRQEVNPVR   QEIESETTSE   EQIQEEKSFE   SPQNSPATEQ
1151   TIFDSQTFTE   TELKTTDYSV   LTTKKTYSDD   KEMKEEDTSL   VNMSTPDPDA
1201   NGLESYTTLP   EATEKSHFFL   ATALVTESIP   AEHVVTDSPI   KKEESTKHFP
1251   KGMRPTIQES   DTELLFSGLG   SGEEVLPTLP   TESVNFTEVE   QINNTLYPHT
1301   SQVESTSSDK   IEDFNRMENV   AKEVGPLVSQ   TDIFEGSGSV   TSTTLIEILS
1351   DTGAEGPTVA   PLPFSTDIGH   PQNQTVRWAE   EIQTSRPQTI   TEQDSNKNSS
1401   TAEINETTTS   STDFLARAYG   FEMAKEFVTS   APKPSDLYYE   PSGEGSGEVD
1451   IVDSFHTSAT   TQATRQESST   TFVSDGSLEK   HPEVPSAKAV   TADGFPTVSV
1501   MLPLHSEQNK   SSPDPTSTLS   NTVSYERSTD   GSFQDRFREF   EDSTLKPNRK
1551   KPTENIIIDL   DKEDKDLILT   ITESTILEIL   PELTSDKNTI   IDIDHTKPVY
1601   EDILGMQTDI   DTEVPSEPHD   SNDESNDDST   QVQEIYEAAV   NLSLTEETFE
1651   GSADVLASYT   QATHDESMTY   EDRSQLDHMG   FHFTTGIPAP   STETELDVLL
1701   PTATSLPIPR   KSATVIPEIE   GIKAEAKALD   DMFESSTLSD   GQAIADQSEI
1751   IPTLGQFERT   QEEYEDKKHA   GPSFQPEFSS   GAEEALVDHT   PYLSIATTHL
1801   MDQSVTEVPD   VMEGSNPPYY   TDTTLAVSTF   AKLSSQTPSS   PLTIYSGSEA
1851   SGHTEIPQPS   ALPGIDVGSS   VMSPQDSFKE   IHVNIEATFK   PSSEEYLHIT
1901   EPPSLSPDTK   LEPSEDDGKP   ELLEEMEASP   TELIAVEGTE   ILQDFQNKTD
1951   GQVSGEAIKM   FPTIKTPEAG   TVITTADEIE   LEGATQWPHS   TSASATYGVE
2001   AGVVPWLSPQ   TSERPTLSSS   PEINPETQAA   LIRGQDSTIA   ASEQQVAARI
2051   LDSNDQATVN   PVEFNTEVAT   PPFSLLETSN   ETDFLIGINE   ESVEGTAIYL
2101   PGPDRCKMNP   CLNGGTCYPT   ETSYVCTCVP   GYSGDQCELD   FDECHSNPCR
2151   NGATCVDGFN   TFRCLCLPSY   VGALCEQDTE   TCDYGWHKFQ   GQCYKYFAHR
2201   RTWDAAEREC   RLQGAHLTSI   LSHEEQMFVN   RVGHDYQWIG   LNDKMFEHDF
2251   RWTDGSTLQY   ENWRPNQPDS   FFSAGEDCVV   IIWHENGQWN   DVPCNYHLTY
2301   TCKKGTVACG   QPPVVENAKT   FGKMKPRYEI   NSLIRYHCKD   GFIQRHLPTI
2351   RCLGNGRWAI   PKITCMNPSA   YQRTYSMKYF   KNSSSAKDNS   INTSKHDHRW
2401   SRRWQESRR
```

Structural and functional sites

Putative signal sequence: 1–20
Hyaluronan-binding domain: 48–346
   Immunoglobulin repeat: 48–148
   Link protein repeats: 149–244, 250–346
Glutamate-rich region: 400–408
Potential N-linked glycosylation sites: 57, 330, 411, 455, 481, 676, 911, 1192, 1285, 1293, 1373, 1398, 1405, 1509, 1641, 1947, 2080, 2382, 2392
EGF repeats: 2103–2140, 2141–2178
Lectin repeat: 2179–2305
Complement control protein repeat: 2306–2366

Gene structure

The human versican gene is found on the long arm of chromosome 5 at locus 5q12–5q14 [5].

References

[1] Mole, J.E. et al (1984) Complete primary structure for the zymogen of human complement factor B. J. Biol. Chem. 259: 3407–3412.

[2] Kristensen, T. and Tack, B.F. (1986) Murine protein-H is comprised of 20 repeating units, 61 amino acids in length. Proc. Natl Acad. Sci. USA 83: 3963–3967.

[3] Krusius, T. et al (1987) A fibroblast chondroitin sulfate proteoglycan core protein contains lectin-like and growth factor-like sequences. J. Biol. Chem. 262: 13120–13125.

[4] Zimmermann, D.R. and Ruoslahti, E. (1989) Multiple domains of the large fibroblast proteoglycan, versican. EMBO J. 8: 2975–2981.

[5] Iozzo, R.V. et al (1992) Mapping of the versican proteoglycan gene (CSPG2) to the long arm of human chromosome 5 (5q12–5q14). Genomics 14: 845–851.

Vitronectin is a cell adhesion and spreading factor found in plasma and the extracellular matrix. The molecule binds to proteins at the terminal stages of both complement and coagulation pathways and inhibits cytolysis. It also interacts with cells through a number of integrin receptors containing the αV subunit, principally $\alpha V \beta 3$. Vitronectin participates in a variety of protective events including haemostasis, phagocytosis, tissue repair and immune function. It is synthesized predominantly in the liver, but platelets, macrophages and smooth muscle cells can produce a similar molecule. Immunofluorescence localization suggests that vitronectin is deposited in a fibrillar form in a number of connective tissues.

Molecular structure

In plasma, vitronectin exists in two forms, a single chain (molecular weight of approximately 75 000) and an endogenously clipped, two-chain form held together by disulphide bonds (molecular weights of approximately 65 000 and 10 000). Vitronectin is an asymmetrically shaped molecule with a large content of predicted β-sheet structure. Conformational transitions may lead to activation of binding sites. The NH_2-terminal somatomedin B domain is an independently folded structural module which precedes the RGD-containing cell-binding domain and a highly acidic domain homologous to hemopexin. Somatomedin B is a growth hormone-dependent serum factor with proteinase-inhibiting activity. This region also binds plasminogen activator inhibitor–1(PAI–1). In the two domains that are homologous to hemopexin, there are only six typical repeats as opposed to eight in hemopexin itself. The two domains are connected by a flexible hinge facilitating ligand-induced conformational changes. A polycationic cluster is located at the COOH-terminal end of the molecule and consitutes the major heparin-binding domain [1-4].

Light chain

Somatomedin B   Heparin/C9/PAI-1

PAI-1 Collagen/   Plasminogen/Perforin

FXIIIa X-linking

Isolation

Vitronectin is readily adsorbed onto a variety of surfaces and can be extracted most conveniently on a column of glass beads or by affinity chromatography with anti-vitronectin antibody or heparin–agarose. Major difficulties encountered relate to its sensitivity towards proteases and denaturation, and its tendency to form disulphide-linked oligomers [5]. The molecules isolated by each procedure may exist in different conformational states.

Accession number

P04004; P01141

Primary structure

Sequence conflicts: 50 C to N
225 S to N
366 A to T
400 T to M

Ala A 34	Cys C 14	Asp D 32	Glu E 34
Phe F 23	Gly G 37	His H 9	Ile I 14
Lys K 20	Leu L 32	Met M 7	Asn N 15
Pro P 37	Gln Q 26	Arg R 36	Ser S 34
Thr T 19	Val V 22	Trp W 10	Tyr Y 23

Mol. wt (calc.) = 54 245   Residues = 478

```
  1    MAPLRPLLIL   ALLAWVALAD   QESCKGRCTE   GFNVDKKCQC   DELCSYYQSC
 51    CTDYTAECKP   QVTRGDVFTM   PEDEYTVYDD   GEEKNNATVH   EQVGGPSLTS
101    DLQAQSKGNP   EQTPVLKPEE   EAPAPEVGAS   KPEGIDSRPE   TLHPGRPQPP
151    AEEELCSGKP   FDAFTDLKNG   SLFAFRGQYC   YELDEKAVRP   GYPKLIRDVW
201    GIEGPIDAAF   TRINCQGKTY   LFKGSQYWRF   EDGVLDPDYP   RNISDGFDGI
251    PDNVDAALAL   PAHSYSGRER   VYFFKGKQYW   EYQFQHQPSQ   EECEGSSLSA
301    VFEHFAMMQR   DSWEDIFELL   FWGRTSAGTR   QPQFISRDWH   GVPGQVDAAM
351    AGRIYISGMA   PRPSLAKKQR   FRHRNRKGYR   SQRGHSRGRN   QNSRRPSRAT
401    WLSLFSSEES   NLGANNYDDY   RMDWLVPATC   EPIQSVFFFS   GDKYYRVNLR
451    TRRVDTVDPP   YPRSIAQYWL   GCPAPGHL
```

Structural and functional sites

Signal peptide: 1–19
Somatomedin B domain: 20–63
Hemopexin repeats: 150–287, 288–478
RGD cell adhesion site: 64–66
Potential N-linked glycosylation sites: 86, 169, 242
Protease cleavage site: 398–399
Intrachain disulphide bond: 293–430
Phosphorylation site for cAMP-dependent protein kinase: 397
Sulphation sites: 75, 78
Factor XIIIa transglutaminase-catalysed cross-linking site: 112
Heparin-binding domain: 362–395

Gene structure

The human vitronectin gene is 4.5–5 kb long, comprises eight exons and generates a 1.7 kb mRNA transcript. It is located in the centromeric region of human chromosome 17q. There is no evidence of alternative splicing [6].

References

[1] Hayman, E.G. et al (1985) Vitronectin – a major cell attachment-promoting protein in foetal bovine serum. Exp. Cell Res. 160: 245–258.
[2] Suzuki, S. et al (1985) Complete amino acid sequence of human vitronectin deduced from cDNA. Similarity of cell attachment sites in vitronectin and fibronectin. J. Biol. Chem. 259: 15307–15314.

[3] Jenne, D. and Stanley, K.K. (1987) Nucleotide sequence and organisation of the human S-protein gene: Repeating motifs in the "pexin" family and a model for their evolution. Biochemistry 26: 6735–6742.

[4] Preissner, K.T. (1991) Structure and biological role of vitronectin. Annu. Rev. Cell Biol. 7: 275–310 .

[5] Yatohgo, T. et al (1988) Novel purification of vitronectin from human plasma by heparin affinity chromatography. Cell Struct. Funct. 13: 281–292.

[6] Fink, T.M. et al (1992) The human vitronectin (complement S-protein) gene maps to the centromeric region of 17q. Human Genetica 88: 569–572.

von Willebrand factor

von Willebrand factor has an important function in the maintenance of haemostasis by promoting platelet–vessel wall interactions at sites of vascular injury. The molecule is a multimeric plasma glycoprotein synthesized by endothelial cells and megakaryotes; it serves as both a carrier for factor VIII and as a mediator of initial platelet adhesion to the subendothelium. Multimerization provides a greater density of platelet-binding sites. von Willebrand factor interacts with two known platelet receptors, the membrane glycoprotein complexes Ib/IX and IIbIIIa (integrin αIIbβ3), as well as with collagen and heparin. IIbIIIa recognition involves a single RGD motif. Ib/IX recognition is induced by the viper venom protein botrocetin.

Molecular structure

In plasma, von Willebrand factor circulates as multimers ranging in size from dimers of about 500 kDa to multimers of 20 000 kDa, at a concentration of 5–10 µg/ml. The molecule is synthesized as a pre-propolypeptide which is cleaved and modified. Its cysteine residues (which account for 8.3% of the total amino acids) are all involved in inter- and intrachain disulphide bonds and are clustered at the NH$_2$- and COOH-termini of the molecule. After translocation, the pro-von Willebrand factor forms dimers which then multimerize in the Golgi as a result of disulphide bonding. The propeptide is cleaved from most, but not all, of the subunits producing the range of multimers that are seen in the circulation. The majority of the protein is made up of four types of repeating module (A–D). The A-type repeats are found in a number of other proteins, including cartilage matrix protein and collagen type VI. The C-type repeats share some similarity with the N-propeptide of procollagen type I (also found in thrombospondin) [1–8].

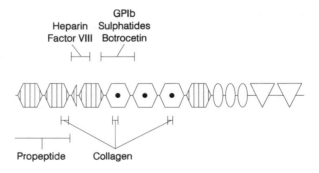

Isolation

von Willebrand factor and factor VIII can be isolated from plasma by cryoprecipitation and agarose gel filtration. It is possible to separate factor VIII from von Willebrand factor by gel filtration in high ionic strength buffers [4].

Accession number

P04275

Primary structure

Sequence conflicts: 484 R to H
770 P to H
804 C to S
1472 D to H
1914 S to T
2168 S to C

Ala A 154	Cys C 233	Asp D 157	Glu E 182	
Phe F 88	Gly G 206	His H 68	Ile I 95	
Lys K 108	Leu L 228	Met M 56	Asn N 98	
Pro P 176	Gln Q 132	Arg R 144	Ser S 208	
Thr T 151	Val V 225	Trp W 26	Tyr Y 78	

Mol. wt (calc.) = 308 913   Residues = 2813

1	MIPARFAGVL	LALALILPGT	LCAEGTRGRS	STARCSLFGS	DFVNTFDGSM
51	YSFAGYCSYL	LAGGCQKRSF	SIIGDFQNGK	RVSLSVYLGE	FFDIHLFVNG
101	TVTQGDQRVS	MPYASKGLYL	ETEAGYYKLS	GEAYGFVARI	DGSGNFQVLL
151	SDRYFNKTCG	LCGNFNIFAE	DDFMTQEGTL	TSDPYDFANS	WALSSGEQWC
201	ERASPPSSSC	NISSGEMQKG	LWEQCQLLKS	TSVFARCHPL	VDPEPFVALC
251	EKTLCECAGG	LECACPALLE	YARTCAQEGM	VLYGWTDHSA	CSPVCPAGME
301	YRQCVSPCAR	TCQSLHINEM	CQERCVDGCS	CPEGQLLDEG	LCVESTECPC
351	VHSGKRYPPG	TSLSRDCNTC	ICRNSQWICS	NEECPGECLV	TGQSHFKSFD
401	NRYFTFSGIC	QYLLARDCQD	HSFSIVIETV	QCADDRDAVC	TRSVTVRLPG
451	LHNSLVKLKH	GAGVAMDGQD	VQLPLLKGDL	RIQRTVTASV	RLSYGEDLQM
501	DWDGRGRLLV	KLSPVYAGKT	CGLCGNYNGN	QGDDFLTPSG	LAEPRVEDFG
551	NAWKLHGDCQ	DLQKQHSDPC	ALNPRMTRFS	EEACAVLTSP	TFEACHRAVS
601	PLPYLRNCRY	DVCSCSDGRE	CLCGALASYA	AACAGRGVRV	AWREPGRCEL
651	NCPKGQVYLQ	CGTPCNLTCR	SLSYPDEECN	EACLEGCFCP	PGLYMDERGD
701	CVPKAQCPCY	YDGEIFQPED	IFSDHHTMCY	CEDGFMHCTM	SGVPGSLLPD
751	AVLSSPLSHR	SKRSLSCRPP	MVKLVCPADN	LRAEGLECTK	TCQNYDLECM
801	SMGCVSGCLC	PPGMVRHENR	CVALERCPCF	HQGKEYAPGE	TVKIGCNTCV
851	CRDRKWNCTD	HVCDATCSTI	GMAHYLTFDG	LKYLFPGECQ	YVLVQDYCGS
901	NPGTFRILVG	NKGCSHPSVK	CKKRVTILVE	GGEIELFDGE	VNVKRPMKDE
951	THFEVVESGR	YIILLLGKAL	SVVWDRHLSI	SVVLKQTYQE	KVCGLCGNFD
1001	GIQNNDLTSS	NLQVEEDPVD	FGNSWKVSSQ	CADTRKVPLD	SSPATCHNNI
1051	MKQTMVDSSC	RILTSDVFQD	CNKLVDPEPY	LDVCIYDTCS	CESIGDCACF
1101	CDTIAAYAHV	CAQHGKVVTW	RTATLCPQSC	EERNLRENGY	ECEWRYNSCA
1151	PACQVTCQHP	EPLACPVQCV	EGCHAHCPPG	KILDELLQTC	VDPEDCPVCE
1201	VAGRRFASGK	KVTLNPSDPE	HCQICHCDVV	NLTCEACQEP	GGLVVPPTDA
1251	PVSPTTLYVE	DISEPPLHDF	YCSRLLDLVF	LLDGSSRLSE	AEFEVLKAFV
1301	VDMMERLRIS	QKWVRVAVVE	YHDGSHAYIG	LKDRKRPSEL	RRIASQVKYA
1351	GSQVASTSEV	LKYTLFQIFS	KIDRPEASRI	ALLLMASQEP	QRMSRNFVRY
1401	VQGLKKKKVI	VIPVGIGPHA	NLKQIRLIEK	QAPENKAFVL	SSVDELEQQR
1451	DEIVSYLCDL	APEAPPPTLP	PDMAQVTVGP	GLLGVSTLGP	KRNSMVLDVA
1501	FVLEGSDKIG	EADFNRSKEF	MEEVIQRMDV	GQDSIHVTVL	QYSYMVTVEY
1551	PFSEAQSKGD	ILQRVREIRY	QGGNRTNTGL	ALRYLSDHSF	LVSQGDREQA
1601	PNLVYMVTGN	PASDEIKRLP	GDIQVVPIGV	GPNANVQELE	RIGWPNAPIL
1651	IQDFETLPRE	APDLVLQRCC	SGEGLQIPTL	SPAPDCSQPL	DVILLLDGSS
1701	SFPASYFDEM	KSFAKAFISK	ANIGPRLTQV	SVLQYGSITT	IDVPWNVVPE
1751	KAHLLSLVDV	MQREGGPSQI	GDALGFAVRY	LTSEMHGARP	GASKAVILV

1801	TDVSVDSVDA	AADAARSNRV	TVFPIGIGDR	YDAAQLRILA	GPAGDSNVVK
1851	LQRIEDLPTM	VTLGNSFLHK	LCSGFVRICM	DEDGNEKRPG	DVWTLPDQCH
1901	TVTCQPDGQT	LLKSHRVNCD	RGLRPSCPNS	QSPVKVEETC	GCRWTCPCVC
1951	TGSSTRHIVT	FDGQNFKLTG	SCSYVLFQNK	EQDLEVILHN	GACSPGARQG
2001	CMKSIEVKHS	ALSVELHSDM	EVTVNGRLVS	VPYVGGNMEV	NVYGAIMHEV
2051	RFNHLGHIFT	FTPQNNEFQL	QLSPKTFASK	TYGLCGICDE	NGANDFMLRD
2101	GTVTTDWKTL	VQEWTVQRPG	QTCQPILEEQ	CLVPDSSHCQ	VLLLPLFAEC
2151	HKVLAPATFY	AICQQDSSHQ	EQVCEVIASY	AHLCRTNGVC	VDWRTPDFCA
2201	MSCPPSLVYN	HCEHGCPRHC	DGNVSSCGDH	PSEGCFCPPD	KVMLEGSCVP
2251	EEACTQCIGE	DGVQHQFLEA	WVPDHQPCQI	CTCLSGRKVN	CTTQPCPTAK
2301	APTCGLCEVA	RLRQNADQCC	PEYECVCDPV	SCDLPPVPHC	ERGLQPTLTN
2351	PGECRPNFTC	ACRKEECKRV	SPPSCPPHRL	PTLRKTQCCD	EYECACNCVN
2401	STVSCPLGYL	ASTATNDCGC	TTTTCLPDKV	CVHRSTIYPV	GQFWEEGCDV
2451	CTCTDMEDAV	MGLRVAQCSQ	KPCEDSCRSG	FTYVLHEGEC	CGRCLPSACE
2501	VVTGSPRGDS	QSSWKSVGSQ	WASPENPCLI	NECVRVKEEV	FIQQRNVSCP
2551	QLEVPVCPSG	FQLSCKTSAC	CPSCRCERME	ACMLNGTVIG	PGKTVMIDVC
2601	TTCRCMVQVG	VISGFKLECR	KTTCNPCPLG	YKEENNTGEC	CGRCLPTACT
2651	IQLRGGQIMT	LKRDETLQDG	CDTHFCKVNE	RGEYFWEKRV	TGCPPFDEHK
2701	CLAEGGKIMK	IPGTCCDTCE	EPECNDITAR	LQYVKVGSCK	SEVEVDIHYC
2751	QGKCASKAMY	SIDINDVQDQ	CSCCSPTRTE	PMQVALHCTN	GSVVYHEVLN
2801	AMECKCSPRK	CSK			

Structural and functional sites
Signal peptide: 1–22
Propeptide: 23–763
A-type repeats: 1260–1479, 1480–1672, 1673–1874
B-type repeats: 2296–2330, 2340–2365, 2375–2399
C-type repeats: 2400–2515, 2544–2662
D-type repeats: 23–295, 363–652, 657–741 (partial), 842–1130, 1934–2203
Potential N-linked glycosylation sites: 857, 1147, 1231, 1248, 1255, 1256, 1263, 1468, 1477, 1486, 1487, 1515, 1574, 1679, 2223, 2290, 2298, 2357, 2400, 2546, 2585, 2790
Sulphation site: 652
RGD cell adhesion site: 2507–2509
GPIb-binding sites: 1237–1251, 1457–1471

Gene structure

The von Willebrand factor gene spans 178 kb on the short arm (p) of human chromosome 12, contains 52 exons and produces an 8.7 kb mRNA. A partial unprocessed von Willebrand factor pseudogene is located on chromosome 22 (locus q11–13); this spans 21–29 kb and corresponds to exons 23–34 of the authentic von Willebrand factor gene [6].

References
[1] Bonthron, D. et al (1986) Nucleotide sequence of pre-pro-von Willebrand factor cDNA. Nucleic Acids Res. 14: 7125–7127.
[2] Verweij, C.L. et al (1986) Full length von Willebrand factor (vWF) cDNA encodes a highly repetitive protein considerably larger than the mature vWF subunit. EMBO J. 5: 1839–1847.

3 Zimmerman, T.S. and Meyer, D. (1987) Structure and function of factor VIII and von Willebrand factor. In: Haemostasis and Thrombosis, 2nd edition, Bloom, A.L. and Thomas, D.P., eds, Churchill Livingstone, Edinburgh, pp. 131–147.

4 Mohri, H. et al (1988) Structure of the von Willebrand factor domain interacting with glycoprotein Ib. J. Biol. Chem. 263: 17901–17904.

5 Mancuso, D.J. et al (1989) Structure of the gene for human von Willebrand factor. J. Biol. Chem. 264: 19514–19527.

6 Sadler, J.E. (1991) Von Willebrand factor. J. Biol. Chem. 266: 22777–22780.

7 Ruggeri, Z.M. and Zimmerman, T.S. (1987) Von Willebrand factor and von Willebrand disease. Blood 70: 895–904.

8 Ginsburg, D. et al (1992) Fine mapping of monoclonal antibody epitopes on human von Willebrand factor using a recombinant peptide. Thrombosis Haemostasis 67: 166–171.

Index